AF674962

PRODUCTION TECHNOLOGY OF LUMP SUGAR *GUR*/JAGGERY

PRODUCTION TECHNOLOGY OF LUMP SUGAR–*GUR*/JAGGERY

AJIT K. GHOSH
ASHOK K. SHRIVASTAVA
VIRENDRA P. AGNIHOTRI

2016
Daya Publishing House®
A Division of
Astral International Pvt. Ltd.
New Delhi - 110 002

© AUTHOR
First Published, 1998
Reprinted, 2016

ISBN: 978-81-7035-188-7 (Hardbound)

Publisher's note:

Every possible effort has been made to ensure that the information contained in this book is accurate at the time of going to press, and the publisher and author cannot accept responsibility for any errors or omissions, however caused. No responsibility for loss or damage occasioned to any person acting, or refraining from action, as a result of the material in this publication can be accepted by the editor, the publisher or the author. The Publisher is not associated with any product or vendor mentioned in the book. The contents of this work are intended to further general scientific research, understanding and discussion only. Readers should consult with a specialist where appropriate.

Every effort has been made to trace the owners of copyright material used in this book, if any. The author and the publisher will be grateful for any omission brought to their notice for acknowledgement in the future editions of the book.

All Rights reserved under International Copyright Conventions. No part of this publication may be reproduced, stored in a retrieval system, or transmitted in any form or by any means, electronic, mechanical, photocopying, recording or otherwise without the prior written consent of the publisher and the copyright owner.

Published by : **Daya Publishing House®**
A Division of
Astral International Pvt. Ltd.
– ISO 9001:2008 Certified Company –
4760-61/23, Ansari Road, Darya Ganj
New Delhi-110 002
Ph. 011-43549197, 23278134
E-mail: info@astralint.com
Website: www.astralint.com

Laser Typesetting : **Classic Computer Services**
Delhi - 110 035

Digitally Printed at : **Replika Press Pvt. Ltd.**

PRINTED IN INDIA

Preface

From times immemorial *gur* has satisfied the human craving for nutritive sweeteners. The lump sugar—*gur* or *jaggery* is a nutritive sweetening agent manufactured primarily from sugarcane. It is consumed almost by all Indians in some form or the other. To date, its per capita consumption in India is 10.6 kg (in 1990-91). In some parts of the world *gur* is also produced from certain palms and sweet sorghum.

Gur or *jaggery* is primarily produced in developing or under developed countries. Considerable developments have taken place on various aspects of *gur* manufacturing process. The present effort critically reviews most of the technological developments in this field of research. It identifies bottlenecks in the *gur* production technology and how it can be improved upon. It highlights the methodologies for making *gur* from sugarcane growing under biotic and abiotic stress conditions.

Different aspects of *gur* making from various palms and sweet sorghum have been provided. Various aspects of the palm *gur* discussed include tapping of the sap and its processing to *gur*, special types of *gur* and syrup obtained from palms and its composition relative to the *gur* obtained from sugarcane juice. In the chapter on *gur* from sweet sorghum, manufacturing of *gur* from juice of sweet sorghum stalks, constraints in its manufacture and how these could be overcome, its quality *vis-a-vis gur* from sugarcane juice, syrup preparation, etc., have been discussed. The poineering work done in this field by the Nimbkar Agricultural Research Institute in Maharashtra finds its due place. Engineering aspects related to crushers and furnaces have been included as these are must for increasing *gur* recovery. Constraints in *gur* manufacturing have also been identified.

Medicinal uses of *gur* as identified by indigenous medical practitioners for keeping the body healthy and strong have been listed.

We strongly feel that if the improved technology for *gur*

making is adopted and research efforts are geared in appropriate direction, in times to come the product will be able to fulfil the increasing nutritive sweetener demands and wider acceptability besides augmenting the employment potential in this unorganised sector.

As this adequately illustrated book deals with various aspects of *gur* from sugarcane, palms and sweet sorghum, we strongly feel that a vademecum of this type will be useful to agronomists, progressive farmers, *gur* manufacturers and those involved in *gur* storage and marketing.

A.K. Ghosh
A.K. Shrivastava
V.P. Agnihotri

Acknowledgements

We wish to express our gratitude to all the scientists and colleagues at the Indian Institute of Sugarcane Research, Lucknow, who have provided valuable helps, in various ways in preparation of this manuscript. We express our gratitude to Shri P.N. Avasthy for critically going through the manuscript and suggesting several improvements. Our sincere thank are due to Dr. N.S.L. Srivastava from Central Institute of Agricultural Engineering, Bhopal for contributing a chapter on sugarcane crushers and Drs. K. Sengupta, S.K. Das and C. Sen, from Bidhan Chandra Krishi Viswavdyalya, Kalyani for their valuable contribution on palm *gur*. We are also grateful to Dr. N.K. Tiwari, Dr. A.C. Srivastava, Shri Ram Darsh and Shri S.P. Shukla for their help in preparing the manuscript.

Authors

Contents

List of Figures

1

INTRODUCTION

From times immemorial craving of human beings for sweeteners, to make their food delicious and palatable, has been satisfied by honey and juice of certain ripe fruits like *figs, jamun,* etc. With the development of civilization many other plants containing sugars in some of their parts have been recognised (Table 1.1).

The three important sources of sugar are sugarcane (*Saccharum* spp. hybrids), sugarbeet (*Beta vulgaris* L.) and sugar maple (*Acer saccharum* Marsh). Other plants have, however, not gained much importance for a variety of reasons like their relatively lower sugar content, problems associated with their large scale cultivation, harvesting, sugar extraction, clarification and geographical distribution. Some of these, however, are still exploited in limited geographical location(s), as minor source(s) of sugar (or sugar in their juice/sap is fermented to intoxicating liquor).

The solid or semi-solid mass that we get after boiling sugarcane juice is called *gur* in north India and *jaggery* in south India. It is also made from the sap obtained by tapping inflorescence or the stalk of certain palms and sweet sorghum. *Gur* or the lump sugar is one of the most ancient sweetening agents known to *Homo sapiens.* Its passing reference has also come in *Rigveda*—the old Indian Sanskrit epic. This lump sugar is known by different names in India. For example, *gur* in Hindi, *gul* in Marathi, *bella* in Kannada, *bellam* in Telegu and in Tamil as well as *guda* in Sanskrit.

Gur industry is one of the most important cottage industries in India from times immemorial. *Gur* is prepared mostly by small and marginal farmers. Besides India, countries like Pakistan, Bangladesh, Nepal, Myanmar (erstwhile Burma) and Philippines also manufacture *gur.* In Ecuador, Bolivia, Peru and Columbia *gur* is called *Chancaca of Panela,* in Brazil as

Table 1.1: Plants containing sugar

Plant		Family	Part containing sugar	Sucrose % juice/sap
Sugarcane species				
	Saccharum officinarum	*Gramineae*	Stalks (juice)	3.25 - 20.90
	S. barberi	*Gramineae*	Stalks (juice)	4.16 - 17.79
	S. robusta	*Gramineae*	Stalks (juice)	9.37 - 10.09
	S. sinensis	*Gramineae*	Stalks (juice)	15.53 - 16.50
	S. spontaneum	*Gramineae*	Stalks (juice)	6.27 - 16.50
Sugarcane	(*Saccharum* spp. hybrids)	*Gramineae*	Stalks (juice)	7.0 - 20.0
Sugarbeet	(*Beta vulgaris* L.)	*Chenopodiaceae*	Roots (juice)	17.0
Sugar maple	(*Acer saccharum* Marsh.)	*Aceraceae*	Sap wood of lower part of the trunk - tapping	1.3 - 3.0
Palms		*Palmae*	Trunk or inflorescence	
(a) Coconut	(*Cocos nucifera* L.)		Crown base	14.0
(b) Palmyra palm	(*Borassus flabellifer* L.)		Spadix or fruit	12.0
(c) Sugar palm	(*Arenga pinnata* (Wurmb.) Merr.)			
(d) Fish tail palm	(*Caryota urens* L.)			
(e) Toddy palm or wild date palm	(*Phoenix sylvestris* Roxb.)		Spadix	10.0
(f) Nipa palm	(*Nypa fruticans*)			
(g) Buri palm	(*Corypha elata*)			
(h) Talipot palm	(*Corypha unbraculifera*)			
Sweet sorghum	(*Sorghum vulgare* Pers. var. *saccharatum* (L.) Boerl.)	*Gramineae*	Stalks (juice)	9.4 - 10.6
Manna	(*Fraxinus ornus* L.) (*Alhaqi naurorun* Medic.)	*Oleaceae* *Leguminosae*	Branches bearing flowers (manna is obtained after slashing these)	
Tamarisk manna	(*Tamarix mannifera* Ehrenbg.)	*Tamaricaceae*	Scale insect (*Coccus manniparus)* either punctures the stem triggering the exudation or secrete the manna itself.	
Figs	(*Ficus carica* L.)	Moraceae	Ripe fruits	
Grape vine	(*Vitis vinifera* L.)	Vitaceae	Ripe fruits	
Jamun	(*Syzygium jambolana*)	Myrtaceae	Ripe fruits	

Source: Shrivastava, 1992; Sreenivasan and Bhagyalakshmi, 1993.

Rapadura, in Mexico as *Pilocillo,* in Venezuela as *Papelon* and in Philippines as *Panocha* or *Muscavado.* Besides sugar and *gur,* sugarcane is also used, to a large extent, for the manufacture of *khandsari* (crystalline sugar produced by open pan process). In international sugar trade these two sweetening agents (*gur* and *khandsari)* are known as non-centrifugal sugars and also popularly referred to as 'artisan sugar'.

The early history of cane sugar could be traced back as early as three thousand years ago. The primitive technology though subject to some variations from place-to-place essentially remained same based on vegetable and other methods of clarifying raw juice and boiling to a solid or semisolid product and it was handed down through generations. Pearson (1989) believed that it originated in the Far East from where it spread across Asia to Egypt onto the Islands of the eastern Mediterranean. The process of juice clarification could be traced *vis-a-vis* the development of chemistry. By early 1700s' when European nations were establishing colonies in the Western Hemisphere and East Indies, there was a uniformity in technique and a growing understanding of the factors which could affect the purity of the product. It was particularly observed that the iuice when left for a long time became acidic and on boiling yielded a dark coloured mass. Lime was generally used to neutralise the acid and bleaching with SO_2 produced favourable affect on *gur* colour.

Victory for the North in the American Civil War outlawed slavery and the sugar producers had no option but to mechanise the sugar production. In 1870s, the industry adopted multiple effect evaporation, the centrifugal machine and steam engines to drive conveyors and lift the juice and create vacuum for low temperature crystallization (Pearson, 1989). By the end of nineteenth century, the crude sugar was no longer accepted in the commercial centres of the world and was relegated to its present status of primary sweetener suitable for small scale, low capital initial input and produced only in the under developed or developing parts of the world (Smith, 1978).

The technical efficiency of open pan sugar (OPS) process is far lower than the vacuum pan sugar (VPS) process because of poor extraction and prolonged processing at higher temperatures. However, the yield on cane basis is of similar order because little or no separation of molasses, the product has a higher moisture content and retains rather more impurities.

In India, the traditional methods to produce a relatively fine grade yellowish-brown crystal sugar—the *khandsari* were modified adopting centrifugation and lime sulphitation for juice clarification. This yielded a crystal sugar nearly approaching the standards of vacuum pan sugar in quality.

Various countries in the world produce nearly 15 million tonnes of non-centrifugal sugars (Table 1.2). It has been observed that at the global level production of OPS is still tending to increase though at a relatively slower pace than the vacuum pan sugar (VPS), and its share of total world sugar supply is declining. The OPS production and trade continue to be an important economic activity in certain regions. Besides, OPS remains an important component of the diet of the people and in some countries it is the principal home-grown sweetener (Kaplinsky, 1989).

In India, there are about 25 million small sugarcane growers crushing about 50 per cent of the sugarcane produced, for making *gur*. The lump sugar, produced by ignored cottage industry, is used for preparing innumerable dishes in different parts of the country. *Gur* is consumed by about 80 per cent of the rural population in India. The production of *gur* has increased from two million tonnes in 1930-31 to over 8.5 million tonnes in 1990-91 (Table 1.3). The yield of *gur* has also increased from 3.6 t ha^{-1} to 5.14 t ha^{-1} during this period.

In India, out of the total sugarcane produced in 1991, 52.1 per cent was utilised for producing white crystal sugar and 36.0 per cent for *gur* and *khandsari* (Fig. 1.1a). In subtropical India, 39.06 per cent of the sugarcane was utilised for producing white crystal sugar and 48.04 per cent for *gur* and *khandsari* (Fig. 1.1 b). On the other hand, in tropical India, 67.16 per cent of the cane was utilised for the production of white crystal sugar and only 22.09 per cent for *gur* and *khandsari* (Fig. 1.1c).

In India the major *gur* markets are Ankapalle, Bhopal, Calcutta, Delhi, Hapur, Kolhapur, Chennai, Patna and *khandsari* markets are at Bareilly and Delhi (Anon., 1992).

Figures of per capita consumption of sweeteners (sugar, *gur* and *khandsari)* as a function of time, indicate that total per

Table 1.2: Production of non-centrifugal sugars in various parts of the world

NCS producing parts of the world	NCS produced (×000 t)			
	1979-81	1987	1988	1989
World (as a whole)	13304	11130	12307	12239
Africa	78	89	89	89
Kenya	30	28	28	27
Nigeria	33	40	40	41
Tanzania	5	6	6	6
Uganda	10	15	15	15
North Central America	255	235	248	243
Costa Rica	17	12	12	12
El Salvador	18	23	23	23
Guatemala	39	34	38	39
Haiti	80	84	84	78
Honduras	21	21	21	21
Mexico	68	51	58	58
Nicaragua	10	11	10	10
Panama	2	2	2	2
South America	1199	1228	1214	1289
Brazil	200	270	240	240
Colombia	925	882	897	972
Eucador	53	54	54	54
Peru	14	13	13	13
Venezuela	7	9	10	10
Asia	11773	9575	10757	10619
Bangladesh	461	420	458	620
China	351	414	448	502
India	8107	6700	8100	7600
Indonesia	30	36	46	41
Japan	10	13	12	12
Myanmar	139	147	150	150
Pakistan	1765	1201	898	1048
Philippines	29	19	19	20
Sri Lanka	—	11	11	12
Thailand	600	600	600	600
Vietnam	11	14	14	15
Europe	—	—	—	—
Oceania	—	—	—	—
Developed Middle East	10	10	10	10

Source: FAO *Production Year-book*, 1989, Vol. 43, pp. 211-212.

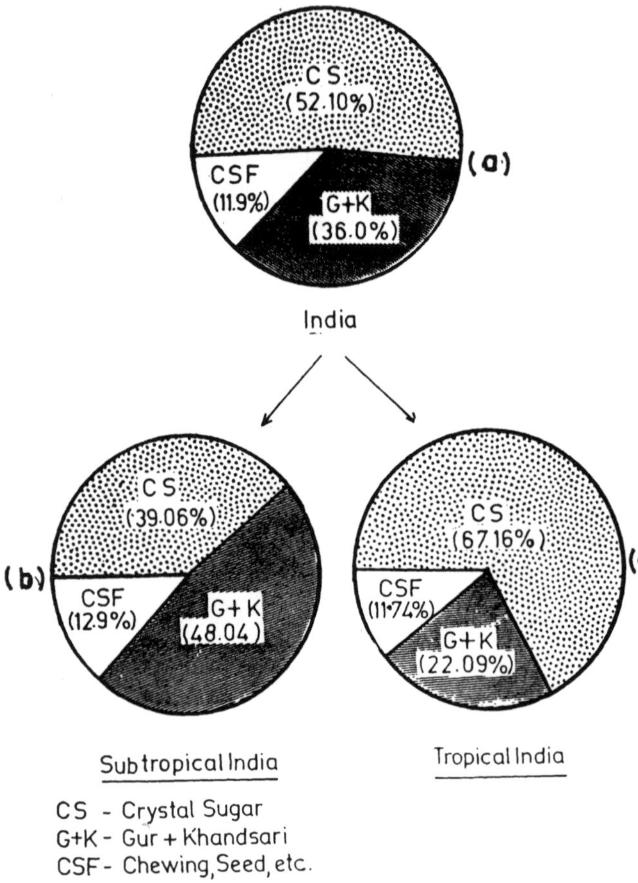

Fig. 1.1 Utilization of Sugarcane: (a) In India, (b) Subtropical India and (c) Tropical India

capita consumption of sweeteners has not appreciably changed and it remained 20.1 ± 1.65 higher over the years from 1960-61 to 1989-90. However, with increase in living standard, per capita sugar consumption has gradually increased from 4.8 kg during 1960-61 to 12.2 kg during 1989-90. Conversely, the *gur* and *khandsari* consumption which was 15.2 kg in 1960-61 has decreased to 10.3 in 1989-90 (Anon., 1992).

Table 1.3: Utilisation of Cane for Sugar, *Gur* and *Khandsari* Production, and their Per Capita Consumption in India

Crop season	Production of *gur* and *khand-sari* (x000 t)	Proportion of cane utilised (%) for		Per capita consumption (kg) of		
		Sugar	*Gur+ Khand-sari*	Sugar	*Gur+ Khandsari*	Total
1984-85	9105	35.3	52.9	10.9	12.4	23.3
1985-86	8284	40.2	48.0	11.1	11.0	22.1
1986-87	7925	45.8	42.3	11.4	10.4	21.8
1987-88	8300	47.8	40.2	11.9	10.7	22.6
1988-89	9391	42.2	45.5	12.1	11.5	23.6
1989-90	8573	49.9	38.2	12.2	10.3	22.5

Source: Co-op. Sugar, 23(10): 704, 1992.

Per capita consumption of sweeteners, and population by 2001 A.D. were estimated using exponential growth curves from the data obtained from 1960-61 to 1985-86 (Anon., 1987). It was estimated that by 2001 A.D. our population will be around 1048 million and per capita consumption of sugar, *gur* and *khandsari* will be 15.84 kg and 10.18 kg, respectively. This predicted population agrees with the population projections published by the Registrar General Census Commissioner of India, in March 1984 which indicated that we may achieve a level of population of 959,991 and 1053 million in low, medium and high series projection groups, respectively (Chaudhary, 1986). To meet the growing demand of sweeteners by 2001 A.D., 285.75 million tonnes of sugarcane will be required.

National Commission on Agriculture (1976) has estimated that by the year 2001 A.D., the population will be around 935.35 million and per capita requirement for sugar will be 14.23 and that of *gur* + *khandsari* 17.2 kg. To meet this requirement about 345 million tonnes of sugarcane will be required (Anon., 1982).

Table 1.4: Composition of *Gur*

Constituent	Percentage
Carbohydrates	
Sucrose	65-80
Reducing sugars	9-12
Inorganic constituents	
Ash	0.5-1.0
Calcium	0.39-0.40
Magnesium	0.08
Chloride	0.34
Sulphate	0.50
Phosphate	0.045
*Iron (mg/100 g gur)	11.06
*Copper (mg/100 g gur)	0.767
Nitrogen	
—Total N (mg/100 g *gur*)	0.57
—Non-protein N (mg/100 g *gur*)	0.33
—Protein N (mg/100 g *gur*)	0.14
#Fats	0.705
Vitamins	
—Vitamin A (μg/100 g *gur*)	280
—Vitamin B (μg/100 g *gur*)	20
—Nicotonic acid (μg/100 g *gur*)	1
pH	5.8-6.4

*Vishwanathan (1942), other values are from *Proc. I.S.S.C.T.*, 9: 558-565.
#Mitra (1940).

Processing of cane for *gur* production has a number of advantages over white crystal sugar. These include: (i) *gur* is manufactured at the place of harvest avoiding its transhipment, (ii) the farmer can harvest the crop according to his convenience, (iii) semi-skilled persons are needed in *gur* industry against skilled persons employed by the sugar industry, (iv) nutritive value of *gur is* more than the white sugar, and (v) *gur* is more suited to our climate from health point of view.

The use of *gur* as a sweetening agent is gradually increasing because it contains appreciable amount of carbohydrates, proteins, minerals and vitamins (Table1.4) which are basic food ingredients for human consumption. Mitra (1940) compared nutritional value of *gur* prepared from various sources with

Table 1.5: Nutritional Value of Gur as Compared to Commercial White Sugar

Particulars	Per cent Composition						
	Protein	Fat	Sucrose	Glucose	Total minerals	Calcium	Phosphate
White sugar	—	—	99.7	—	—	—	—
Gur from cane	0.25	0.05	50.71	21.18	3.36	0.40	0.045
Gur from palmyra	1.04	0.19	70.86	1.66	3.15	0.861	0.052
Gur from date palm	1.46	0.26	72.01	1.48	2.60	0.369	0.052
Gur from coconut palm	0.96	0.15	71.89	3.70	5.04	1.638	0.062
Gur from sago palm	2.28	0.11	84.31	0.53	3.66	1.352	0.372

Source: Mitra, 1940.

commercial white crystal sugar (Table 1.5). Nutritional composition indicates that palm *gur* is relatively a better nutritional source than the *gur* prepared from sugarcane.

Gur has an important bearing on the development of sugarcane industry and cultivation in subtropical India. Barber (1919) suggested "in the past comparative lack of communications and the presence of hardy indigenous kinds; the need of sugar or its equivalent *gur* caused the sugarcane to become an integral part of cultivation in north India, whereas the smaller populations of the south were accessible from out side and had other sources of supply in their forests". He also observed that "the principles of gur making are undoubtedly in many places woefully wasteful". Similar view was also expressed earlier by Mr. Sayer (*c.f.* Barber, 1919). Barber (1919) further emphasised the need of an organised investigation into the local practices of *gur* making throughout the country, with the object of introducing improvements according to local conditions.

In many *gur* producing areas of the country, concerted research efforts were made to improve this important cottage industry. Technology has been developed to prepare *gur* from lodged, pest and disease ridden, waterlogged, and stale canes and to store it for a relatively longer time in hot and humid weather. Significant improvements have been made in the designs of crushers and furnaces to profitably sustain the *gur* industry. Blanching of sugarcane at 80-90°C has improved *gur* yields from stale/discarded canes. Methods have been developed to judge the *gur* making ability of sugarcane varieties obviating actual boiling of juice. The open pan sugar (OPS) process, as a sugar processing technology, emerged as a craft skill in India in the late fifties.

Preparation of special types of *gur* from sugarcane juice

In the market, special types of *gur* are available and for their manufacture particular type of clarificant is required. Manufacture of such type of *gur* has been suggested for presenting the produce to the consumers in different forms and varieties (Anon., 1955; 1961).

Semi Solid Gur

For the preparation of semi-solid *gur* alum is used; it is added after the juice has been clarified with one or other clarificant (as mentioned in Chapter 7). Alum, being acidic in nature, inverts the sucrose in juice at the time of boiling. As a result, it does not allow the *gur* to set well. This brilliant light colour semi-solid *gur* is devoid of crystalline structure.

This type of semi-solid *gur* is liked by the people of Assam, Bengal and Bihar.

Neutral Gur

For the preparation of this type of *gur*, sugarcane juice is clarified by sucrate of lime. It is prepared by first slaking quick-lime with a small quantity of water. Additional quantity of water is added after the lime has been completely slaked to bring the density to approximately 20°Be. About 18 parts of milk of lime is added to 100 parts of juice and the mixture is stirred vigorously to settle the colloidal matter. The clear supernatant liquid is called the 'sucrate of lime'.

Fresh sucrate of lime is added to the juice after it has been clarified by using usual clarificants. Lime sucrate is added till the pH of the juice becomes almost neutral (6.8-7.0 pH). The pH of the juice is checked by B.T.B. indicator paper; it turns from yellow to blue.

After neutralising the juice, it is boiled again and scums frequently removed by using a perforated laddle. The scum free juice is then concentrated to make *gur* using the customary method.

Gur prepared in this way is hard and has crystalline structure. Due to the neutralisation of free organic acids, the keeping quality of the *gur* is excellent. Although the *gur* is slightly dark in colour, its *net rendement* (recoverable sugar) is higher; this type of *gur* is specially suitable for *gur* refineries. However, it is not advisable to manufacture this *gur* for direct consumption as beside being dark in colour it is also not palatable.

Cream Jaggery

This term is coined for *gur* manufactured after decolourising the cane juice with activated vegetable carbon (paddy husk carbon). The *gur* has a very attractive cream colour and is highly palatable because it retains original flavour and taste, as no foreign matter is left in the finished product.

For preparation of cream jaggery, cane juice is first boiled with vegetable clarificant and scum is removed from time to time. When scum production almost stops, then activated carbon is added to the juice (1% on the weight of juice). The juice boiling is continued for 10 more minutes and scums appearing on the surface of the juice are again thoroughly removed. The hot juice is then transferred to another pan placed at a higher level and siphoned off into a bucket filled with a layer of fresh activated carbon underlaid with sand and small pebbles. The bucket has a spout at the bottom to allow the carbon treated juice to pass out after traversing through the layers of materials in the bucket. During this process the juice gets, by and large, decolourised. Carbon particles, if present, are carefully removed by sprinkling diluted milk (1 part of milk : 8 parts of water) over the juice and the scum formed is removed using a perforated laddle covered with a wire gauze of fine mesh. When there is no scum formation, the juice looks almost colourless in the boiling pan. This colourless juice is then boiled to prepare *gur* by the usual method. *Gur* prepared by this method retains its very light cream colour as also its original flavour for a long time. However, two things come in the way of producing this type of *gur*. Firstly, recovery of cream jaggery on cane weight basis is slightly lower because of repeated clarification of the juice. Secondly, the cost of production is higher due to the high cost of the activated carbon and the additional labour required. In order to reduce the cost of *gur* preparation and to manufacture cream jaggery on a commercial scale, settling tanks may be used instead of bucket percolaters, as mentioned above (Anon., 1955; 1961).

Powdered gur (Shakkar)

This is practically a special type of brown sugar meant for direct consumption. This type of *gur* is particularly liked by inhabitants of western Uttar Pradesh. They prefer to eat it as such, with pure *ghee* or with pure *ghee* and black pepper. For the preparation of this *gur* same methods of clarification and boiling are employed, as in the case of ordinary *gur* except that boiling in the final stage is continued a little longer and the mass is struck from pan at a temperature of 120°C . It is then allowed to cool in the boiling pan, a little amount of *soda*, dissolved in water, is sprinkled on the hot *gur* mass when it has cooled a little. The mass is then kneaded vigorously with scrappers. Thereafter, it is made into heap and allowed to cool further for about 15 minutes. Then the mass is rubbed vigorously between the hands till a powdered *gur* is produced. Addition of a little sodium hydrosulphite along with *soda* makes such *gur* a little lighter in colour to resemble powder sugar.

Kakavi

In Kolhapur (Maharashtra), one of the largest *gur* marketing centres in India, liquid *gur* is also manufactured on a large scale. Here, the crystallization of the *gur* is arrested at the strike point by adding citric acid. The liquid *gur*, called *kakavi*, is bottled, preserved and used as and when needed.

REFERENCES

Anonymous (1955). Clarificants used in the manufacture of *gur. I.I.S.R. Newsl.* 1(1-2): 13-16.

Anonymous (1961). Manufacture and Storage of Gur in the U.P., Indian Central Sugarcane Committee, Ministry of Food and Agriculture, Government of India, New Delhi, p. 15.

Anonymous (1982). *Indian Agriculture in Brief,* 9th edition, Directorate of Economics and Statistics, Deptt. Agril. and Co-operation, Min. Agril. and Irrigation, New Delhi, pp. 501.

Anonymous (1987). *Co-op. Sugar* 19(3): 202.

Anonymous (1992). *Co-op. Sugar* 23(10): 704.

Barber, C.A. (1919). Progress of the Sugarcane Industry in India during the Years 1916 and 1917. Agricultural Research Institute, Pusa *(Bihar), Bulletin No. 83,* pp. 45, Supdt., Govt. Printing, India, Calcutta.

Choudhary, M. (1986). Population policy target of net reproduction rate of unity by 2001 A.D.: How realistic. *Econ. & Political Weekly,* 21 (48): 210.

Kaplinsky, R. (1989). Cane Sugar - The Small-Scale Processing Option, Proceedings of Jt. ITDG/IDS Conference, Sept., 10-11, 1987, Intermediate Technology Publication, London, pp. 229.

Mitra, K. (1940). Food value of cane sugar and *gur (jaggery). Patna J. Medicine* 14(4).

Pearson, J.H. (1989). History and development of the technology, in Cane Sugar - The Small-Scale Processing Option, (Ed. Kaplinsky, R.), Proceedings of Jt. ITDG/IDS Conference, Sept., 10-11, 1987, Intermediate Technology Publication, London, pp. 31-33.

Shrivastava, A.K. (1992). Sugar maple. *Sugar Crops Newsl.* (I.I.S.R., Lucknow, India), 2(1): 2-3.

Sreenivasan, T.V. and K.V. Bhagyalashmi (1993). Varietal improvement for increasing sugar production, presented at the National Symposium on Improvement on Sugarcane Quality for Increasing Sugar Production, Indian Institute of Sugarcane Research Lucknow, India, Sept 21-23.

Smith, D. (1978). *Cane Sugar World*, Palmer Publications, pp. 328.

Vishwanathan, R.B.B. (1942). Comparative food value of *gur (jaggery)*, sugar and molasses. *The Market Review*, pp. 44-47.

2

HISTORY OF *GUR*

Manufacture of Sweeteners from Sugarcane in Ancient China

Literature indicates that China was one of the earliest countries where the cane was grown and sugar was made. The Chinese cane sugar industry was flourishing in the southern regions of China, when Marco Polo, an Italian, visited China during 1270-1275. He referred frequently to the elegance of their sweet products and the esteem in which they were served. However, the origin was well known before this. The first record of sugar dates back to the second century B.C., and production methods were known before the Chin and Hyan Dynasties (225 B.C. to 200 A.D.). Many old Chinese books referred to by Chang (1992) mentioned that sugar was first made as a syrup, then as a solid, and finally as a granulated product. Refining was practised in the seventh century A.D. In 100 B.C., Yang Fu, a Cantonese, wrote a book *Yi Wu Che* in which he mentioned 'stone honey' a solidified product made from concentrated syrup from sugarcane. It was similar to hard stone but as sweet as honey. Tao Hung-Tsin (452-536 A.D.) wrote a book *Min Yi Pi Lu* in which he also mentioned the healthy use of granulated cane sugar. A great treatise on Chinese agriculture *Chin Min Yao Shu* written by Chia Sze-Shih (533-544) during the Northern Wei Dynasty, discussed briefly the varieties of cane grown and methods of production. In the period of the Northern Sung Dynasty (960-1126), Wang Shao, a native of Shi Ning in Szechuan Province, wrote a book *Tang Shuang Pu* meaning *A Treatise on Sugar Frost* in which origin of sugarcane varieties, crushing cane to extract juice and the methods of producing crystallized sugar were summarised. Probably the sugar made was similar to frost in appearance. This was also perhaps the first book consisting of seven chapters exclusively devoted to sugarcane and sugar (Chang, 1992). He has given extract from the book as follows:

> Some varieties of cane were good only for chewing, while others were good for sugar manufacture. Sugarcane was peeled, cut into small pieces and milled to extract juice. The mill was made of two round stones, one above the other. The upper one, being bullock drawn part was 7 or 8 feet long weighing more than one thousand pounds. The lower stone served as a base, with a circumference more than ten feet. In other cases a mortar was used instead, and the cane was pounded. The sugarcane, in pieces about the size of a coin, were extracted through the stone mills, then pressed in bamboo press. The juice was received in lacquered jars. The juice was boiled to a proper concentration and the strips of bamboo were then thrown into the jar containing it. Two days later, sugar began to crystallize out and small crystals could be seen and tested with fingers. Gradually small crystals appeared on the end of the bamboo strips, increasing in size similar to peas or fingers, or forming groups of crystals like a hill. The product was finally dried in the sun.

Later on a great treatise on agricultural and industrial technology *Tien Kung Kai Wu* was written by Sung Ying-Hsing in 1637, during Ming Dynasty (*c.f.* Chang, 1992). This was a fairly big book, comprising eighteen volumes. The sixth volume is on sweets, dealing with the varieties of cane, sugar products, sugar manufacture, white sugar, animal sugar, honey and maltose. Chang (1992) has illustrated the parts relevant to sugarcane and sugar as given in the book mentioned above. The cane juice was extracted in a primitive wooden crusher consisting of two horizontal platforms; at the four corners were inserted wooden pillars, well buried in the ground to stabilise the crusher. At the centre of the upper platform there were two holes, holding two deeply corrugated rolls of hard wood. The one protruding through the upper platform was connected to a piece of bent wood, driven by an ox. In this way, one of the rolls was in motion, while the other was stationary. The two toothed rolls were in perfect coordination, so that the sugarcane passing through was well crushed, while the juice drained off into a jar through a trough under the crusher. The cane residue was crushed again to extract more juice. The juice was largely extracted after three crushings.

A proper amount of lime was added for clarification and scum floating on the top was skimmed off. The clean juice was boiled in three pots, arranged in series, one for heavy juice and the others for light juice. The fire was controlled by watching the state of boiling.

When the heavy juice was boiled to the proper concentration, it sprinkled like peas, became viscous and could be tested with fingers. The concentrated juice was then emptied into tanks for cooling. After cooling, the mixture of crystallized sugar and molasses was poured into cones, placed over big receiver pots, with holes at the bottom plugged with straw. The cones served as crystallizers. When the crystallization was complete, the straw plugs were taken away and molasses allowed to drain off through the holes into the pots. A piece of heavy paper was placed on top of the sugar, and covered with a layer of clay paste. The water in the clay paste drained through the sugar below, washing away the remaining molasses, leaving very white sugar in the upper part of the cone, but slightly brownish yellow underneath. The whiteness of the sugar was superior to that made in foreign countries.

For producing candy sugar, after skimming off the dirt, the clear sugar liquor was evaporated to proper concentration. Strips of bamboo were thrown in the liquor before it was allowed to cool overnight. Large crystals as candy sugar were thus obtained.

This primitive method of producing cane sugar remained in use for centuries throughout the sugar industry in China until sugar factories using more modern machinery were built in the middle of twentieth century (Chang, 1992).

Buddhism and Sugarcane

Buddhism appears to be intimately related with the sugar manufacture and Buddhist literature (in Pali language) provides useful information regarding history of this sweetening agent. Daniels and Daniels (1976) have mentioned the development of sugar technology in India and China. They have mentioned that *Vinaya Pitaka* (i.e., Rules of discipline) is the first section of *Tripitaka*. *Patimokkha* which is the nucleus of *Vinaya Pitaka* enumerates 227 rules for regulation of the *Sangha*, established by Buddha during his life time. This dates back to 500 B.C. There is a mention of *Phanita* (syrup) as one of the five medicines used to treat sick monks. In those days monks were not allowed to drink fermented liquor or spirits. Under spirits, comes five *asavas* or extracts, one is from sugar. This indicates that sugar was known at least in 500 B.C. for its secondary processing to alcohol. *Vinaya Pitaka* also mentions the advent of a new sugar product and some thing similar to a sugar factory.

The new sugar product *gula* (= *guda* in Sanskrit) completely replaced *Phanita* (syrup) which appears exclusively in the earlier *Patimokkha* section. *Gula*, a generic name for raw sugar in its primitive form, is *Phanita* boiled dry into an amorphous mass. With developing technology, *Phanita* was crystallized and concentrated until it could be rolled into balls; this product was also known as *gula*. As no syrup was removed, the difference was only in texture and not in purity. This change in name indicates some improvement in technology from the time when *Patimokkha* was formulated (500 B.C.) to when the histories were codified (300 B.C.)

By the time of *Charaka*, 1st -2nd century A.D., at least five sugar products were known. These were *phanita, guda, matsyandi, khanda* and *sarkara*. The last three products involved the removal of impurities, in varying amounts from syrup; the purest form being *sarkara* (washed white). These three products were not mentioned in *Vinaya Pitaka* and indicated that the technology involved was developed after 300 B.C. There had been a mention in *Vinaya Pitaka* that in sugar factories people were mixing flour with syrup to make sugar so as to make it firm and the resultant product was referred to as sugar. Watt (1893) mentioned about the use of ashes of certain plants for clarification of cane juice in 19th century in India; and rice flour for refining sugar in China. The latter was also used in 618-906 A.D. to make a hard slab or 'lump type' of sugar.

In Chinese *Vinaya*, sugarcane is usually translated *kan che* (sweet bamboo) and *Phanita* as *shi mi* (stone honey). There is rather no distinction between *gula* (raw sugar) and *Phanita* (syrup) in the Chinese *Vinaya*. When sugar cane juice is left for a few days under the sun, it forms *i* (soft or liquid sugar prepared by saccharification of cereal starch). Some people also called it *shi mi*. Barley sugar in its hard form was known as *hsing* and can also be decorated and dried in the sun. Once solidified, it is like ice and when broken its surface is glossy (Daniels and Daniels, 1976).

By 200 B.C. in China, sugar cane syrups, *che chiang* and *che t'ang* were produced. Production of amorphous raw sugar, *shi mi,* commenced in Annam in 3rd century A.D. by sun drying. There were further developments in the technology and with simple clarification a higher grade of *shi mi* was achieved by 4th century. The art of producing crystalline raw sugar *sha'ang* was a later development (about 500 A.D.), and probably introduced from India.

Historical facts indicate that the Chinese Emperor, T'ai Tsung in 647 A.D. sent a mission to Magadha (India) to learn the art of sugar manufacture. Perhaps this is the first instance, on record, of a technological commission investigating a manufacturing process in a foreign country (Ghosh, 1947).

In ancient Indian literature, we get some mention of *gur.* In the district Pratabgarh three important sugarcane varieties viz., *Saroti, Kuswar* and *Katara* dominated in those days. Only the first two varieties yielded *gur* of excellent quality. The best quality *gur* was obtained from *paragana* Patti where one *bigha* of good cane, as a rule, produced 15 *maunds* (5.55 qtl.) of *gur.* About 14 *seers* (12 kg) of *gur* was sold for one rupee. In Unnao, people were under the impression that thin canes (*Bara unkha*) yield more of saccharine matter than thicker canes (*Barangha* and *matra).* The secret, however, lies in their crude mill which is nothing more than the sharpened end of a small tree, working with a rotary motion inside a partially hollowed log, being unable to express the juice from a finer kind of cane, while the inferior one readily yields to the pressure brought to bear upon it (Anon., 1878-79).

Non-Centrifugal Sugars

Non-centrifugal sugars, one-seventh of the world's sugar supply, are seldom recognised or taken into account in either production or consumption statistics periodically issued by the London-based International Sugar Organisation. Many countries which make non-centrifugal sugars seem to give little attention to it, probably due to the similarity of the manufacturing process to the one used about a century ago for all commercial sugars. The manufacturing of this product involves skill on the part of the maker, rather than the science involved in the manufacture of centrifugal sugars. For this reason, non-centrifugal sugar is widely referred to as *artisan sugar.* World production of non-centrifugal sugars (brown lump sugar) was 12,000,000 t and most of it was produced in Asia. About 170,000 t of Palm sugar, in lumps, is produced in Asian tropical countries from the juice of coconut, palmyrah and date palm. Maple sugar or syrup or lump is produced in Canada and northern United States from the sap of sugar maple trees. The sap from the later is obtained by tapping the lower part of the trees in early spring (Anon., 1978).

Artisan sugar is recognised by different names and is manufactured in different forms which vary from country to country. The main centres of production of such products are located in Latin America, Asia including India. In western Hemisphere, nearly every country produces significant quantities of artisan sugar; Colombia being the largest producer. The process of manufacture is almost identical in Mexico, Peru, Brazil, etc. The amount of artisan sugar produced in Latin American countries during 1975-76 is given in Table 2.1.

Table 2.1: Production of Artisan Sugar in Latin American Countries (1975-76)

(In 1,000 metric tonnes)

Argentina	10	Guatemala	54
Bolivia	15	Haiti	15
Brazil	200	Mexico	65
Columbia	687	Nicaragua	10
Costa Rica	40	Panama	8
Dominion Republic	8	Peru	13
Equador	40	Venezuela	38
El Salvador	16		
		Total	1.219

Source: Estimate by Smith (1978).

Artisan sugar is known as *panela* in Western Hemisphere, *pincillo* in Mexico, *papelon* in Venezuela, *banque* or *rapadura* in Brazil, *rapadou* in Haiti, *panochas* or *muscavado* in Philippines and *panela* or *chancaca* in Colombia, Bolivia and Peru, etc. Irrespective of the name of artisan sugar given by countries of the western Hemisphere, good quality *panela* is usually of loaf shape and golden in colour. It contains most of the reducing sugars and other soluble constituents present in sugarcane. Poor quality artisan sugar (produced by inferior and unripened cane) is dark in colour and gum like in nature similar to *rapadou* prepared in Haiti. In some cases, *panela* is manufactured to the tune of 6-8 thousand tonnes per year. Substantial quantity of total sugarcane produced is converted into non-centrifugal sugar, in Thailand. The product is mostly consumed by rural population. But it is sold both in rural and urban (city) markets. Small farmers process the sugar and sell it in the market as *nam tarm sai daeng* or red sugar. In Japan, farther south in Kagoshima perfecture, a black coloured non-centrifugal sugar is produced. The inhabitants of this island have preference for such type of product due to its taste. In Nepal, however, 20-25 per cent of the total production of sugarcane is converted into *jaggery.* South Vietnam has a number of artisan sugar mills of moderate size which annually produce approximately 4,000 tonnes of *yellow sugar.* The sugar has a bright yellow colour similar to *panela* produced in central South America and sells at a premium over the white sugar in Saigon and nearby areas. In North, Central and South Vietnam about 8,000 to 10,000 tonnes of artisan sugar is produced. Two types of sugar are produced, cane sugar and a special large crystal product. In cone, the top part is relatively light and the colour depends on the bottom part. This is sold in the market and merchants cut the cone into sections and thereby produce six different grades of sugar. These artisan sugar cones are used in ceremonial functions and fetch relatively high price. During manufacture, specially large crystals, three to four cm in length, are formed. About five per cent of Mexican sugar is a non-centrifugal product known as *pilocillo or panela.* It is mostly produced in central and southern part of the country. The product is marketed within a radius of few miles from the place of its manufacture. In remote areas of Brazil, non-centrifugal sugar (*rapadura*) is produced. The output of this product is in the range of 175,000 to 200,000 mt/year. One-tenth of the sugar consumption in Venezuela is artisan or non-centrifugal sugar which is known as *papelon*. This type of sugar is manufactured in the Andes mountain area of western Venezuela. This sugar is mostly sold in the area where it is produced; the per capita consumption of raw sugar is about 40 kg. *Gur*, the traditional sugar of Bangladesh, is manufactured in rural areas and it is consumed at the place of its production. Sarker *et al.* (1985) have mentioned that more than 50 per cent of total sugarcane produced is utilized for *gur* manufacture. Artisan or non centrifugal sugar in Myanmar (earstwhile Burma) is produced from both sugarcane and palms. The materials made from sugarcane is known as *Kyantaga;* it is manufactured by more or less the same process employed in other south-east Asian countries for the manufacture of raw sugar. The product made from palm is known as palm *gur*. It is somewhat similar to the *gur* produced in India, but it is prepared by the process followed for palm sugar manufacture in Thailand.

Chanaca is non-centrifugal sugar produced from cane juice in Peru by defacation, concentrating and open pan boiling. In this process all the non-sucrose components present in the sugarcane juice are not removed from the raw sugar (Tateo, 1978).

In addition to centrifugal sugar, Indonesia has always produced, like many other developing countries, a certain quantity of open-pan non-centrifugal sugar called *gula djawa* or *gula mangkok* and is primitively made wholesome brown sugar. Its production was 350,000 tonnes in 1969 and 1970 and its cost is usually lower than the centrifugal sugar (Goedhart, 1982).

In Mexico, crude non-centrifugal sugars, consisting of sucrose crystals and molasses moulded into blocks before drying, are produced, usually by the primitive procedure (Smith *et al.*, 1975). In Mexico, *pilocillo* syrup is made. For this crude non-centrifugal sugar is melted in steel tank fitted with agitator and steam coils. When solubilization of sugar is complete, agitation is stopped to permit the inside matter to settle and a clear syrup is decanted for use in boiling operations (Smith *et al.*, 1975).

REFERENCES

Anonymous (1877-78). *Gazetteer of Oudh (1877-78)* , reprinted in 1985, B.R. Publication Corporation, Delhi, Vol. III, pp. 80: 357-520.

Anonymous (1978). World production of non-centrifugal sugar. *Encyclopaedia Britannica (Macropaedia)*, 17: 775.

Chang, Lie-Tien (1992). Cane sugar manufacture in ancient China. *International Sugar J.* 94 (1123): 155-156.

Daniels, J. and C.A. Daniels, (1976). Buddhism, sugar and sugarcane. *Sugarcane Breeder's Newsl.* No. 38: 35-60.

Ghosh, A.K. (147). Sugarcane and Industry in India, *Sci. and Culture* 12: 466-478.

Goedhart, A. (1982). The Indonesian sugar industry: Past, present and future. *Sugar Y Azucar* 77 (12): 23-27.

Sarkar, M.A.A., M.A. Jabber, R.J. Alam and J. Haque (1985). Sugarcane variety for *gur* production. *Bangladesh J. Sugarcane* 7: 76-78.

Smith, B.A., R.V. Romo, R.C. Smith, R.A. dela Cruz, J.J. Molina, R. Rivera and B.J. Line (1975). Production of centrifugal raw sugars from Mexican *pilancillos. Sugar Y Azucar* 70 (13): 25-27.

Smith, D. (1978). *Cane Sugar World, Palmer* Publications, U.K., pp. 328.

Tateo, F. (1978). Production technology of CHANCACA sugar of the INCAS. *Ind. Aliment (Pinerolo, Itlay)* 17: 28-32 (in Italian).

Watt. G. (1893). *The Dictionary of the Economic Products of India.*

3

SUGARCANE VARIETIES *VIS-A-VIS* *GUR* PRODUCTION

Sugarcane varieties have a pivotal role in the production of quality and quantity of *gur* per unit area and time. Even in earlier times in Pratabgarh (Oudh Province), amongst the then prevailing sugarcane (all *Saccharum officinarum*) varieties *Saroti, Kuswar* and *Katara,* the latter was used for chewing whereas the former two were used for *gur* making (Anon., 1877-78). Barnes (1918) has described the *gur* making ability of sugarcane varieties prevailing in the earlier part of this century in the erstwhile Punjab Province of undivided India (Table 3.1).

It is a common experience with *gur* manufacturers to find a wide variation in the quality of *gur* resulting from different varieties although grown under identical conditions. The colour of the produce may vary from yellow to dark brown, texture from crystalline to amorphous and taste from fine to disagreeable. It is true that all varieties suitable for *gur* making can be accepted by the mills, but the reverse is not true (Sirur *et al.* 1953). The reason why some sugarcane varieties yield good quality *gur,* while certain others do not under similar conditions of soil, growth and management is a most important point. This is largely ascribed to variability in chemical constituents other than sucrose content. For the manufacture of *gur,* preference is given to those varieties having juice of high degree of claribility which depends on low levels of colloids, minerals (N, Cl etc.), non-sugars and non-protein nitrogen, high level of phosphate and wider ratio of sucrose to glucose (Khanna and Chacravarti, 1951; Dakshindas and Kale, 1962; Nagrajan *et al.* 1985; Asokan *et al.,* 1991), low value of N : P ratio, K, Ca, Mg, Cl and high content of P, Zn, Mn and Cu (Kapur *et al.,* 1988).

Table 3.1: *Gur* making Efficiency of Sugarcane Varieties at Gurdaspur (Punjab) (1911-12 to 1915-16)

Variety	Ripening behaviour	Gur yield (t/ha)	Remarks
Katha (suitable for *barani* and *sailab* (flooded) conditions)	Early	2.7	*Gur* from *Katha* possess better keeping quality
Dhaulu (require better soil and liberal water supply	Ripens later than *Katha*	3.2	*Gur* is better in colour & sugar made from it is whiter than *Katha*, but is relatively inferior in keeping quality
Kahu (prefers good and mild season)	Very late	2.2-3.6	Produces *gur* and *Shakker* of good quality
Kansar (prefers good land and water supply)	Late	2.6	Yields excellent *gur*

Source: Barnes,1918.

Early ripening and high sugared varieties like Co 62174, Co 6806, Co 7314, Co 7704 and Co C 671 are suitable both for white sugar and *gur* manufacture. However, there are some varieties like Co 775, Co 975, Co 62175 and Co 6907 which are most suitable for *gur* making. In view of the importance of *gur* in our rural economy, systematic varietal selection has been going on at the Sugarcane Breeding Institute, Coimbatore. Attempts are being made to combine good cane yield and *gur* quality. According to Thuljaram Rao *et al*, (1983) a variety exclusively meant for *gur* should have the following characteristics : (i) softness and low fibre content, (ii) medium in stalk diameter, (iii) light coloured and easily clarifiable juice, (iv) low mineral content, and (v) high sucrose per cent with low invert sugars. Excellent quality of *gur* has been prepared by some of the short duration varieties even at 240 days (Anon., 1987).

Considerable variation in *Saccharum* spp. exists in fibre content, sucrose and non-sugars (organic and inorganic) that interfere in the manufacture of either white/raw sugar or *gur*. Although outstanding success has been achieved in producing superior sugarcane genotypes possessing higher yield and tolerance to pests and diseases, no attempt has been made to look into technological aspects of the varieties evolved though it has been realised that many of the milling and clarification problems can be solved to a greater extent by choosing a suitable variety (Rao and Gupta, 1987).

Studies on the characteristics of sugarcane varieties in relation to quality of *gur* indicated that juice quality in relation to *gur* quality was characterized by high sucrose, purity, low reducing sugars, low reducing sugar/sucrose ratio and low amount of amino acid N (Gopala Aiyer and Chiranjivi Rao, 1959; Gopala Aiyer and Krishnamoorthy, 1954 and Anon., 1987). Gopala Aiyer and Krishnamoorthy (1954) reported that good quality *gur* had greater amount of pore space, higher purity and lower reducing sugar content. Gopala Aiyer and Chiranjivi Rao (1959) reported that high quality *gur* was obtained from juice with low amino acid nitrogen. Mohan Rao *et al.* (1951) found that *gur* prepared from Co 527 was superior to *gur* produced from other prevalent varieties of sugarcane. Co 655, Co 6806, Co 6304 and Co 419 in south India and Co J 64 and Co J 67 in north India are better suited varieties for producing *gur* (Table 3.2).

Table 3.2: Performance of Sugarcane Varieties vis-a-vis Gur Production

Variety	Gur production* (per ha)	Variety	Gur recovery** (% cane)
Co 647	5.22	CoJ 76	9.87
Co 650	4.71	Co J 78	9.81
Co 655	5.78	Co J 64	10.47
Co 658	4.62	Co J 79	9.99
Co 419	5.62	Co J 67	10.00
Co 475	5.47	Co J 77	9.71
#Co 6806(E)	7.41	Co J 81	9.63
#Co 1336(E)	5.50	Co 1148	10.20
#Co 510(E)	4.74		
#Co 6304(L)	9.20		

*Data after Sirur *et al.*, 1953.
**Data after Kapur *et al.*, 1988.
#Data after Thangavelu *et al.*, 1984.
E : Early ripening variety; L : Late ripening variety.

Gur prepared from various sugarcane varieties differs in its keeping quality. In a study at Padegaon (Maharashtra) *gur* prepared from eight sugarcane varieties viz. Co 740, Co 775, Co 62059, Co 62085, Co 62175, Co 62222, Co 62227 and P 12666 was stored on wooden rack without any treatment or covering which would effect its keeping quality. It was observed that the *gur* prepared from Co 62222 maintained highest sucrose and lowest reducing sugar contents. Minimum change in colour and loss in weight during storage were observed in the *gur* prepared from Co 62085 and Co 62227, respectively. The lowest sucrose content and highest reducing sugars were obtained in the *gur* prepared from Co 62227 (Gaur *et al.,* 1979).

In north Bihar, sugarcane varieties BO 17, BO 34 and BO 43 are known for their superior *gur* quality besides being resistant to red rot and endowed with high yield potential of 90 to 95 t ha^{-1} and good ratooning ability (Singh *et al.,* 1987).

Table 3.3: Sugarcane Varieties Identified at SRS Anakapalle for Jaggery Making

Variety (Year of release)	Yield (t/ha)			Time of ripening	Resistance to biotic and abiotic stress
	Cane			Gur	Sugar
Co 419 (1936)	120.9	14.08	14.71	March-May (12-14 m)	S-red rot
Co 527 (1946)	93.8	10.08	11.19	Nov.-Jan. (9-10 m)	S-red rot, smut
Co 997 (1959)	97.9	11.96	13.20	Nov.-Jan. (9-10 m)	S-red rot R-drought
Co 975 (1960)	120.9	13.94	16.04	Feb.-Mar. (11-13 m)	R-red rot, withstands waterlogging
Co 62175 (1968)	142.5	17.44	19.18	March-May (12-14 m)	High yielder
Co A 71-1 (1970)	103.2	11.57	14.02	Nov. -Jan. (9-10 m)	S-scale insect
Co A 7602 (1976)	94.82	12.27	12.88	Feb.-March (11-13 m)	R-red rot
Co A 7701 (1977)	94.07	10.79	11.76	Nov.-Jan. (9-10 m)	R-drought
(1979)	115.22	13.53	15.37	Oct.-Dec. (10-12 m)	High yielder

S = susceptible to, R-resistant to

Source: modified from Chiranjeevi, 1987.

Table 3.4: Chemical and Physical Characteristics of *Gur* of Different Varieties

Variety	Sucrose %	Reducing sugars %	Total non-sugars %	Colour of N/2 *gur* solution	Hardness under 5 kg	Moisture %	Gur recovery %
Early ripening varieties							
79 A 28	87.76	3.63	3.53	102	0.31	2.77	12.20
Co 7508	87.29	3.85	3.63	115	0.30	2.94	11.64
Co 6907	84.75	4.66	4.21	120	0.46	3.59	11.05
CoA 8401	83.96	5.19	5.62	132	0.47	3.39	12.68
Mid-season ripening varieties							
CoT 8201	86.34	4.76	4.62	113	0.34	3.19	12.84
Co 7219	86.13	4.68	4.81	118	0.39	3.13	12.32
Late ripening varieties							
CoR 8001	89.55	3.75	4.88	111	0.34	3.20	11.53
Co 7706	97.15	4.25	4.39	108	0.34	3.12	10.93
CoA 8402	84.72	4.18	5.02	114	0.34	3.12	11.60
Co 419	82.53	5.82	5.69	127	0.44	3.43	12.32
Co 62175	83.31	5.31	5.21	128	0.50	3.64	10.33

Source: Anon., (Anakapalle, India) (1989). 75 Years of Agricultural Research at Anakapalle (1913-1988), RARS, Anakapalle, Andhra Pradesh, p. 252).

In Andhra Pradesh, 60 per cent of the cane produced is used for *gur* production. Sugarcane variety Co 419 (evolved in early 1930s and released in 1936) revolutionised *gur* as well as sugar industry. Sugarcane Research Station, Anakapalle has been identifying sugarcane varieties for making *gur* from October to May giving 12 to 14 per cent recovery on cane weight basis. Besides some of these exhibited explicit resistance to certain biotic and abiotic stresses (Table 3.3).

Chemical and physical properties of the *gur* prepared from some quality *gur* producing early, mid-season and late ripening varieties (at Anakapalle, Andhra Pradesh) are given in Table 3.4.

Sugarcane varieties recommended for making *gur* in subtropical and tropical parts of India are mentioned below:

Sub-tropical India

1. Uttar Pradesh

(a) By the end of 19th century: *Saroti, Kuswar* (in Pratabgarh of Oudh Province) #
(b) Earlier part of the 20th century: *Saretha, Chin, J 33, Dhour, Kagzi, Agoul@; Reora of Benaras$*
(c) After 1956:

Co varieties: Co 312, Co 421, Co 453, Co 845, Co 851, Co 859, Co 951, Co 975, Co 1007, Co 1111, Co 1148, Co 1158, Co 1336, Co 6425, Co 6611, Co 6613, Co 6812

CoS varieties: CoS 245, CoS 321, CoS 443, CoS 510, CoS 524, CoS 575, CoS 611, CoS 671, CoS 673, CoS 718, CoS 729, CoS 770, CoS 771

BO varieties: BO 17, BO 32, BO 47, BO 54, BO 70

Sugarcane varieties recommended for *gur* and *khandsari* in various zones of U.P. during 1983 are:

Eastern U.P.: CoS 510, CoS 611, CoS 718, CoS 758, CoS 770, CoS 771, Co 1148, Co 118, BO 17, BO 70

Central U.P.: CoS 611, CoS 718, CoS 767, CoS 802, Co 1148, BO 17, BO 70, BO 91

Western U.P.: CoS 611, CoS 718, CoS 770, CoS 771, Co 1148, Co 1158, Co 6613, Co 6812, BO 54, BO 70

Southern U.P.: CoS 611, CoS 718, CoS 729, CoS 770, Co 1148, Co 1158, BO 70

After 1991, recommended sugarcane varieties for *gur* are :

CoS 687, CoS 767, CoS 802, CoS 837, CoS 7918, CoS 8118, CoS 8119, CoS 8315, Co 1148, BO 91 and U.P. 5.*

Other promising varieties for *gur* production are : (a) in early ripening group: CoS 8436, CoS 88230, CoS 90265

(b) in mid-late ripening group: CoS 8408, CoS 8432, CoS 87231, CoS 86218, CoS 88216, CoS 8016, CoS 90269*

References : Agarwal *et al.* (1985); #Anon. (1877-78); Anon., (Shahjahanpur, India), 1983; Anon., (Coimbatore, India), 1981; $ Clarke *et al.* (1912); @ Hulme and Sanghi (1918); Singh *et al.* (1978); * Singh and Lal (1991); Srivastava *et al.* (1991).

2. *Bihar*

Co 313, Co 508, Co 513

BO 17, BO 34, BO 43, BO 47, BO 70, BO 90, BO 91, BO 102, BO 104, BO 108, BO 109, CoS 767, Co 1148

References: Anon., (Coimbatore, India), 1981; Khanna and Chacravarti (1951; 1955); Singh *et al.* (1987)

3. *Punjab*

(a) Earlier part of the 20th Century: *Kansar, Kahu, Dhaulu and Katha#*

(b) Co 313

Later studies indicated that varieties giving

(i) Best quality of *gur:* CoJ 64, CoJ 76
(ii) Good quality of *gur:* CoJ 78, CoJ 67, CoJ 81
(iii) Medium quality of *gur:* CoJ 79, Co 1148
(iv) Poor quality of *gur:* CoJ 77

References : Anon., (Coimbatore, India), 1981; #Barnes (1918); Sharma *et al.,* (1985).

4. *Haryana*

Before 1966: Co L 9
After 1966: Co 975, Co 1148, Co 1158, Co 7314, Co 7717
References: Chaudhary and Sarial (1987); Taneja *et al.* (1986).

5. *Rajasthan*

Co 313
References: Anon., (Coimbatore, India), 1981.

6. *West Bengal*

Co 527, Co 997, Co 1148, Co 1158, Co 7224, Co 62101, Co 62033, Co J 64

References: Anon., (Coimbatore, India), 1981; Roy (1991).

7. *Madhya Pradesh*

(a) In the early part of the 20th century:

(i) for irrigated sugarcane (in order of *gur* quality): *Khajla, Munhtora, Bhelsai, Mandkia*
(ii) for unirrigated sugarcane: *Ledu** and *Dhaul** gave good quality but *Kansia** gave poor quality *gur*
Co 419, Co 1305

References: #Hadi (1929); Nahatkar *et al.* (1991); *local names.

8. Assam

Co 313

Reference: Anon., (Coimbatore, India), 1981.

Tropical India

1. Andhra Pradesh

(a) Before 1936: J 247, P.O.J. 2878, B. 208*
(b) After 1936: Co 419, Co 527, Co 975, Co 1012, CoA 71-1, CoA 7601, CoA 7602, CoA 7701

Later studies indicated that for quality *gur*:

(i) Early ripening varieties: 79 A 28, Co 7508, Co 6907, CoA 8401
(ii) Mid-season varieties: CoT 8201, Co 7219
(iii) Late ripening vareities: CoR 8001, Co 7706, CoA 8402, Co 419, Co 62175

References: Anon., (Coimbatore, India), 1981; Anon., (Anakapalle, India), 1989; Bhupala Rao *et al.* (1991); #Chiranjeevi.(1987), Reddy *et al.* (1984).

2. Gujarat

Co 77, Co 62175

References: Anon., (Coimbatore, India), 1981; Patel (1987).

3. Maharashtra

Co 475, Co 650, Co 655, Co 67, Co 708, Co 775, Co 1163, Co 7527, Co 7634, Co 7704, Co 8014, Co 209, Co M 7125, Co M 7129, Co 62222

References: Anon. (Coimbatore, India), 1981; Dagade and Randive (1980); Dakshindas and Kale (1962); Gaur *et al.* (1979); Hasabnis *et al.* (1991); Randive (1985) Tembhekar (1954); Wandre *et al.* (1985); Zende, (1984).

4. Karnataka

HM 320, HM 645, HM 661

***Reference:* Hunsigi (1987).**

5. Tamil Nadu

(a) Before 1936: J 247, P.O.J. 2878, B. 208.
(b) After 1936: Co 419, Co 421, Co 475, Co 658, Co 740, Co 1253, Co 6304, Co 6806, Co 62174, Co 7201, Co 7313, Co 7327, Co C 671.
***References:* Anon. (Coimbatore, India), 1981; Anon., (Coimbatore, India) 1974; 1976; 1977; #Chiranjeevi (1987); Gopala Aiyer and Krishnamurthy (1954); Gopala Aiyer and Chiranjivi Rao (1959); Shakunthala (1980-81).**

REFERENCES

Agarwal, M.L., U. Lal, R.K. Sharma and S.K. Verma (1987). Effect of variety, standards of cultivation and juice clarification on *gur* quality parameters. *Bharatiya Sugar,* 12(11): 33-36.

Anonymous (1877-78). *Gazetteer of Oudh (1877-78)* Vol. III, reprinted in 1985 by B.R. Publication Corpn., Delhi, pp. 609.

Anonymous (Anakapalle, India) (1989). *75 Years of Agriculture Research at Anakapalle: 19:3-1989,* Regional Research Station, Anakapalle, APAU, pp. 400.

Anonymous (Shahjahanpur, India) (1983). *70 Years of Sugarcane Research (1912-1981),* U.P. Council of Sugarcane Research, Shahjahanpur, Uttar Pradesh.

Anonymous (Coimbatore, India). *Annl. Rep.,* Sugarcane Breeding Institute, Coimbatore, 1974, 1976, 1977.

Anonymous (Coimbatore, India) (1981). *Varietal Improvement in Sugarcane in India,* Sugarcane Breeding Institute, Coimbatore, India, pp. 25.

Asokan, S., K. Ballah and R. Nagrajan (1991). The relationship between juice constituents and jaggery quality parameters in certain genotypes of sugarcane, presented at the *National Seminar on Increasing Productivity, Quality and Storability of Jaggery (Gur),* (Nov., 30-Dec., 1) organised by Indian Institute of Sugarcane Research, Lucknow and Indian Society of Agricultural Engineers, New Delhi, at the Indian Institute of Sugarcane Research, Lucknow, India.

Barnes, J.H. (1918). Sugar and the sugarcane in the Gurdaspur District. *Agricultural Research Institute Bulletin No. 69,* Supdt., Govt. Printing, India, Calcutta, pp. 100.

Bhupala Rao, J.V.R., V. Raja Bapa Rao, T. Anuradha and S. Ramakrishna Rao

(1991). Sugarcane varietal performance in plant and ratoon cross in relation to jaggery quality, presented at the *National Seminar on Increasing Productivity, Quality and Storability of Jaggery (Gur)*, (Nov., 30-Dec., 1), organised by Indian Institute of Sugarcane Research, Lucknow and Indian Society of Agricultural Engineers, New Delhi, at the Indian Institute of Sugarcane Research, Lucknow, India.

Chaudhary, B.S. and A.K. Sarial (1987). Coimbatore varieties of sugarcane in Haryana. *Platinum Jubilee Souvenir*, Vol. I. Sugarcane Breeding Institute, Coimbatore, India, pp. 63-65.

Chiranjeevi, V. (1987). Exploitation of sugarcane varieties for jaggery production in Andhra Pradesh. *Platinum Jubilee Souvenir* vol. I, Sugarcane Breeding Institute, Coimbatore, India, pp. 66-68.

Clarke, G., H.E. Annett, S.Z. Hussain, S.C. Banerji and N. Hussain (1912). Experiments on the cultivation of sugarcane at the Pratabgarh Experimental Station 1909-1991, Agricultural Research Institute Pusa (Bihar), Supdt. Govt. Printing, India, Calcutta, pp. 29.

Dakshindas, D.G. and R.A. Kale (1962). Studies on factors affecting *gur* quality. *Indian J. Sugarcane Res. and Devpt.*, 6: 1-11.

Gaur, S.L., S.J. Randive and D.V. Javalekar (1979). Effect of some varieties on the keeping quality of *gur*. *Indian Sugar Crops J.*, 6(1): 18-19.

Gopala Aiyer, K.V. and A.V. Krishnamurthy (1954). Juice characteristics of sugarcane varieties and their relation to quality of jaggery. *Proc. Bien. Conf. S.R.D.W.I.*, 2(II): 564-571.

Gopala Aiyer, K.V. and K. Chiranjivi Rao (1959). Studies on the relation of chemical composition of cane juice and the quality of jaggery with special reference to the new Co canes. *Proc. ISSCT*, 10: 825-829.

Gupta, D.N. (1972). Co 336—A sugarcane variety specially suitable for *gur* production in western U.P. *Sugar News*, 4(6): 11.

Hadi, S.M. (1929). The Indian Sugar Industry, Govt. of Bhopal, C.I., printed by F.W. Petch Thacker and Co., Ltd., Bombay, published by Khan Bahadur S.M. Hadi, Director Agriculture, Bhopal State, pp. 289 + vi.

Hasabnis, A.B., S.S. Wandre and A.G. Shinde (1991). Screening of sugarcane varieties harvested at different stages for jaggery quality, presented at the *National Seminar on Increasing Productivity, Quality and Storability of Jaggery (Gur)*, (Nov., 30-Dec., 1), organised by Indian Institute of Sugarcane Research, Lucknow and Indian Society of Agricultural Engineers, New Delhi, at the Indian Institute of Sugarcane Research, Lucknow, India.

Hulme, W. and R.P. Sanghi (1918). The improvement of the indigenous methods of *gur* and sugar making in United Provinces, Agricultural Research Institute Pusa (Bihar), *Bull. No. 82* , Supdt., Govt. Printing, India, Calcutta, pp. 22.

Hunsigi, G. (1987). Varietal needs of Karnataka, *Platinum Jubilee Souvenir*, Vol. I, Sugarcane Breeding Institute, Coimbatore, India, pp. 69-78.

Kapur, J., K.P. Sharma, B. Kanchan and R.S. Kanwar (1988). Evaluation of new promising sugarcane genotypes for *gur* quality. *Bharatiya Sugar*, 13(8): 69.

Kapur, J., K. Gandhi and R.S. Kanwar (1988). Inorganic constituents in cane juice from promising varieties and their influence on the quality of jaggery. *Proc. Annl. Conv. D.S.T.A.*, 38(1): A-243-A246.

Khanna, K.L. and A.S. Chacravarti (1951). Improvement of *gur* industry in India—A brief account of researches conducted in Bihar and lines of future work to be followed. *Proc. Bien. Conf. S.R.D.W.I.*, 1(II): 74-96.

Khanna, K.L. and A.S. Chacravarti (1955). Scientific Monograph-Researches on Technical Aspects Relating to Improvement of *Gur* Industry in Bihar, Indian Central Sugarcane Committee, New Delhi.

Mohan Rao, N.V., R.L. Narsimham and D. Anji Reddy (1951). Studies in quality of jaggery. *Proc. Bien. Conf. S.R.D.W.I.*, 1(II): 143-151.

Nagarajan, R., S. Asokan and K. Mohan Naidu (1985). A note on breeding sugarcane varieties for high quality and recovery of jaggery. *Proc. National Seminar-cum-Group Discussion on Jaggery Manufacture and Storage* (Dec., 18-19), Indian Institute of Sugarcane Research, Lucknow, India, pp 43-46.

Nahatkar, S.B., R.M. Atri and S.R. Sharma (1991). Economics of jaggery production in Madhya Pradesh, presented at the *National Seminar on Increasing Productivity, Quality and Storability of Jaggery (Gur)*, (Nov., 30-Dec. 1), organised by Indian Institute of Sugarcane Research, Lucknow and Indian Society of Agricultural Engineers, New Delhi, at the Indian Institute of Sugarcane Research, Lucknow, India.

Patel, H.S. (1987). Coimbatore varieties in Gujarat. *Platinum Jubilee Souvenir* vol. I, Sugarcane Breeding Institute, Coimbatore, India, pp. 98-104.

Ramakrishna Rao. S., K. Veerbhadra Rao, S.V.R. Bhupala Rao, M. Ramo Rao and V. Raja Bapa Rao (1991). Better sugarcane varieties for jaggery making, presented at the *National Seminar on Increasing Productivity, Quality and Storability of Jaggery (Gur)*, (Nov., 30-Dec., 1), Organised by Indian Institute of Sugarcane Research, Lucknow and Indian Society of Agricultural Engineers, New Delhi, at the Indian Institute of Sugarcane Research, Lucknow, India.

Randive, S.J. (1985). Review of the work done on *gur* at Sugarcane Research Station, Padegaon, *Proc. National Seminar-cum-Group Discussion on Jaggery Manufacture and Storage* (Dec., 18-19), Indian Institute of Sugarcane Research, Lucknow, India, pp. 99-104.

Reddy, M.S., P. Nageshwara Rao and P. Shivalingam (1984). Studies in improving yield and quality of jaggery through varietal selection. *Maharashtra Sugar*, 9(10): 43-45.

Roy, T.K. (1991). Problems and prospects of *gur* industry in West Bengal, presented at the *National Seminar on Increasing Productivity, Quality and Storability of Jaggery (Gur)*, (Nov., 30-Dec., 1), organised by Indian Institute of Sugarcane Research, Lucknow and Indian Society of Agricultural Engineers, New Delhi, at the Indian Institute of Sugarcane Research, Lucknow, India.

Shakunthala, S.V. (1981). *Manual of Sugarcane Production Technology*, Department of Agriculture, Tamil Nadu, pp. 142-143.

Sharma, K.P., R.S. Kanwar and J. Kapoor (1983). Co J 64—An excellent sugarcane variety for jaggery making. *Maharashtra Sugar*, 8(13): 55.

Sharma, K.P., J. Kapur and R.S. Kanwar (1985). Studies on cane juice and *gur* quality of promising varieties. *Proc. National Seminar-cum-Group Discussion on Jaggery Manufacture and Storage* (Dec., 18-19), Indian Institute of Sugarcane Research, Lucknow, India, pp. 61-66.

Singh, M., K.M. Bharadwaj and M.L. Agarwal (1978). Storage of *gur*—A Review. *Cooperative Sugar,* 9(11): 559.

Singh, J.R.P., B.P. Sahi and R.A. Pandey (1987). Varietal improvement in Bihar. *Platinum Jubilee Souvenir,* Vol. I: Sugarcane Breeding Institute, Coimbatore India, pp. 126-129.

Singh, S. and U. Lal (1991). Evaluation of sugarcane varieties for jaggery (*gur*) making, presented at the *National Seminar on Increasing Productivity, Quality and Storability of Jaggery (Gur),* (Nov., 30-Dec., 1), organised by Indian Institute of Sugarcane Research, Lucknow and Indian Society of Agricultural Engineers, New Delhi, at the Indian Institute of Sugarcane Research, Lucknow, India.

Sirur, S.S., M.V. Divekar and V.V. Tembhekar (1953). Promising sugarcane varieties in relation to their performance and *gur* quality. *Proc. Annl. Conv. D.S.T.A.,* 11(I): 158-163.

Srivastva, B.L., T.C. Ramana Rao and B. Baboo (1991). Sugarcane varieties in relation to jaggery (*gur*) production, presented at the *National Seminar on Increasing Productivity, Quality and Storability of Jaggery (Gur),* (Nov., 30-Dec., 1), organised by Indian Institute of Sugarcane Research, Lucknow and Indian Society of Agricultural Engineers, New Delhi, at the Indian Institute of Sugarcane Research, Lucknow, India.

Taneja, A.D., S.N.L. Srivastava, M.S. Punia and S.S. Verma (1986). The study of physicochemical factors for grading of gur of some sugarcane varieties in Haryana. *Bharatiya Sugar,* 11(7): 47.

Tembhekar, V.V. (1954). Effect of mineral ingredients present in the sugarcane juice on the quality of *gur. Proc. Annl. Conv.. D.S.T.A.* 11 (II): 161-163.

Thangavelu, S$_v$ E. Lalitha and K.L. Rao (1984). Jaggery quality of some early and late maturing varieties of sugarcane. *Proc. D.S.T.A., India,* 34: A125-135.

Thangavelu, S. and K.C. Rao (1985). Jaggery quality of early maturing sugarcane varieties at nine months. *Indian J. agric. Sci.,* 55(9): 560.

Thuljaram Rao, J., B.V. Natrajan and K.V. Bhagyalakshmi (1983). *Sugarcane,* Indian Council of Agricultural Research, New Delhi, pp. 140.

Wandre, S.S., A.G. Shinde and D.G. Gojare (1985). Screening of different sugarcane varieties for jaggery and storability. *Proc. National Seminar-cum-Group Discussion on Jaggery Manufacture and Storage* (Dec., 18-19), Indian Institute of Sugarcane Research Lucknow, India, pp. 51-56.

Zende, G.K. (1984). Soil fertility management for higher sugar and sugarcane productivity. *Maharashtra Sugar,* 9(11): 39.

4

QUALITY OF RAW MATERIAL (CANE JUICE) AND ITS EFFECT ON *GUR*

In the early part of this century, it was realised that it is ripeness of the cane at harvest which determined the production of dry and light coloured *gur* (Barnes, 1918). Satyanarayana (1939-43) showed that besides sucrose the proportion of N to the total salts also determined the quality of juice. Work done at Padegaon (Anon., (Padegaon, India), 1933-37) indicated that good quality *gur* is associated with low N, low ash content and high P content in juice. Soluble N and Cl in juice that pass into *gur* are responsible for the deterioration of colour and keeping quality (Gopala Aiyer, 1954). The ratio of sucrose to organic non-sugars in juice is more important than the ash content in affecting the crystalline structure and colour of *gur* (Satyanarayana, 1939-43). The oxidative enzymes present in the buds, nodes and top of the cane react with the phenols present therein to impart brownish colour to *gur* (Anon., 1946). Very high nitrogen and low P content in immature cane is responsible for bad quality, while high ratio of albuminous N to total N in juice contributes to good quality *gur.* Chacravarti *et al.* (1954) observed that low levels of colloids, gums, proteins and ash coupled with high P content of juice were responsible for good quality *gur.*

Accumulation of higher amount of chloride ions under saline conditions was responsible for bad quality *gur* (Krishnamurthy, 1919; Gopala Aiyer and Krishnamurty, 1954). Iron-phenol reaction is the main cause for dark coloured juice (Leley *et al.*, 1955; Pajne and Singh 1963). The anthocyanin and tannins react with the salts present in the juice during boiling and produce dark coloured substances resulting in dark coloured *gur* (Anon., (Shahjahanpur, India), 1965; 1983).

Quality of Raw Material *vis-a-vis* Juice Quality

Ripeness of the cane at the time of harvest is important for making dry and light coloured *gur* (Barnes, 1918). Harvesting premature or over-mature crop is not conducive for production of *gur* and its quality; immature crop contains less sugars and more non-sugars while in over-matured crop inversion takes place (Anon., (Shahjahanpur, India), 1983; Chockalingam, 1985).

In the earlier part of this century, it was a common practice among the farmers of Gurdaspur (Punjab) to allow a day lapse between harvesting and crushing for making *gur.* This practice is said to yield better *gur* with low moisture content (Barnes, 1918). Experience gained in the erstwhile Bhopal State indicated that there were more losses in sugar when stale canes were crushed for making *gur* (Hadi, 1929). It is interesting to note that the quality of *gur* from a *barani* (Unirrigated/rainfed) crop is, as a rule, superior to that from an irrigated crop. In the low lands along rivers and canal banks and on the border of swamps; where absence of drainage prevents other varieties to grow except *katha,* only *rab* is generally made as the crop never properly ripens. In the cane growing areas near Beas, a relatively inferior type of wet sugar called *leb* is produced. It may be mentioned that during those days sugarcane variety *katha* dominated both in *barani* and *sailab* (flooded) lands (Barnes, 1918).

In the erstwhile Bhopal State, sugarcane varieties *Munhtora, Mankia (or Mandkia), Bhelsai (or Bhelsari)* and *Kahajla* were usually grown under irrigated conditions. *Mankia* generally failed in the years of excessive rainfall and the *gur* it yielded was wanting in right flavour, owing to high percentage of salts in the juice. Varieties grown in dry tract of Malwa region were *Kansia, Barru, Ledu* and *Dhaul.* The latter two varieties yielded granular *gur* but the one obtained from *Kansia* was of poor quality (Hadi, 1929).

Immature and deteriorated canes do not yield a good quality *gur.* Studies conducted at Anakapalle, (Andhra Pradesh) indicated that total acidity and non-sugars in the juice extracted from immature and deteriorated canes were higher than in the juice from mature canes (Table 4.1).

In order to improve the *gur* of immature or deteriorated

cane juice, its pH has to be adjusted. Studies at Anakapalle (Andhra Pradesh) indicated that by adopting following methods, it is possible to obtain *gur* of favourably good colour and consistence:

(a) for the juice from immature canes (10 months age) - apply lime (calcium oxide) alone or lime sucrate to pH 6.4.

(b) for the juice from mature canes (12 months age) - apply lime alone or lime sucrate to pH 5.8.

(c) for the juice from deteriorated canes (over-stands 14 months age) - apply lime alone or lime sucrate to pH 6.6, add some *gur* powder at settling stage (Anon., (Anakapalle, India), 1989).

Table 4.1: Composition of Juice from Immature, Mature and Deteriorated Canes (Variety: Co 419)

Particulars	Juice from		
	Immature cane	Mature cane	Deteriorated cane
pH	5.30	5.30	5.10
Acidity (ml 0.1 N NaOH/100 ml)	12.30	10.35	12.63
Sucrose per cent	13.10	17.16	16.05
Purity coefficient	81.92	91.80	87.28
Glucose per cent	1.64	0.90	1.49
Glucose ratio	13.77	5.46	9.33
Ash per cent	0.25	0.27	0.21
Colloids per cent	0.29	0.75	0.53
Total N (mg/100 g)	24.59	15.85	15.09
Protein N (mg/100 g)	17.31	12.15	12.09
Non-protein N (mg/100 g)	7.30	3.70	6.07
Ammoniacal N (mg/100 g)	4.63	3.40	1.87
Amide N (mg/100 g)	1.07	1.93	0.86
Total P_2O_5 (mg/100 g)	25.96	50.29	45.36
Organic P_2O_5 (mg/100 g)	11.62	13.42	25.96
Inorganic P_2O_5 (mg/100 g)	14.41	19.44	18.73

Source: Anon., (Anakapalle, India), 1989.

Addition of some *gur* powder to deteriorated juice just before striking stage appeared to be necessary. It served as a primer or nucleus around which crystallization occurred.

Quality of the *gur* prepared from immature and deteriorated cane may also be improved by blanching cane with hot water (at 80 to 90°C for 20 min) immediately before crushing (Anon., (Anakapalle, India), 1989).

The pest and disease ridden cane did not yield good quality *gur*. When the cane of *Paunda* variety badly affected with red rot was crushed immediately after harvest, it produced *gur* of fairly good quality but when the canes were kept overnight and crushed next morning and the juice boiled into *gur* the final product failed to solidify (Hadi, 1929).

Under swamp conditions, high amounts of non-protein N, organic non-sugars and lower phosphate uptake adversely affect the quality of *gur* manufactured. This *gur* absorbed more moisture when kept at 75 per cent relative humidity (RH) compared to the *gur* made from normal canes (Mohan Rao and Anji Reddy, 1951). Whenever the quality of *gur* was bad (due to variety, excessive nitrogen and swampy conditions), the juice from which it was manufactured had high organic non-sugars, high total and non-protein N, high colloids, high pectins and low amount of P compared to juices yielding better quality of *gur* (Table 4.2) (Mohan Rao *et al.*, 1951; Anon., (Anakapalle, India), 1989).

Table 4.2: Composition of Cane Juice from Swampy Conditions (Variety Co 419)

Particular	Swamp crop (Jan. harvest)	Normal crop (March harvest)
Acidity (ml 0.1 N NaOH/100 ml juice)	10.50	11.00
Total solids (%)	19.20	20.82
Sucrose (%)	16.96	18.82
Glucose (%)	0.90	0.87
Glucose ratio	5.31	4.64
Purity coefficient	88.33	90.39
Total non-sugars (%)	1.34	1.13
Ash (g/100 ml juice)	0.39	0.35
Total N (mg/100 ml juice)	37.25	19.45
Protein N (mg/100 ml juice)	11.52	6.75
Non-protein N (mg/100 ml juice)	25.74	12.70
Non-protein N/Total N (%)	67.09	65.30
P_2O_5 (mg/100 ml juice)	18.60	29.70
P_2O_5 (per cent of ash)	4.60	8.49

Source: Anon., (Anakapalle, India), 1989.

Non-sugar Impurities and Their Effect on Juice Clarification, *Gur* Colour and Storability

Quality of *gur* and its storability depend upon the amount of non-sugar impurities (Tables 4.1 and 4.2) which are partly removed from the juice during clarification. The amount of non-sugars varies between 5-6 per cent in juice. The non-sugar impurities may be both organic (0.5-1%) and inorganic (0.2-0.6%). The colloidal impurities are removed by coagulation by heating the juice. But non-sugar impurities, which are not heat coagulable, remain in juice as lipophilic colloids. There is a need to develop more efficient methods for their removal.

Organic Non-sugars

Natural pigments like chlorophyll, xanthophyll, carotenes, anthocyanin and saccharetin are generally present in juice in varying amounts. Anthocyanin and saccharetin react with salts present in juice and iron of crusher and boiling pan. When juice comes in contact with salts and iron, dark coloured complexes are formed before, during and after clarification; their presence in juice is not desirable (Panje and Singh, 1963).

Proteins

Proteins are largely heat coagulable albuminoids which remain in juice and are broken down to a considerable extent to amino acids under high temperatures during boiling (Khanna and Chacravarti, 1955). To some extent, these proteins are helpful in the clarification of juice. On the other hand, decomposition of products such as proteins, amino acids form complexes with reducing sugars adversely affecting *gur* colour.

Amino Acids and Amides

In sugarcane juice ten amino acids have been identified (Honig, 1953; Wiggins and William, 1951). These are aspartic acid, glutamic acid, leucine, valine, alanine, glycine, serine, lysine, tyrosine and gama-amino butyric acid. These amino acids remain stable in the juice during clarification. Some of the amino acids dissociate during clarification and form complexes with calcium ions and thus their concentration is reduced. Hence, addition of lime is recommended for the clarification of juice. Asparagine and glutamine, the two amides present in the juice, are relatively stable over a wide range of pH but get decomposed at 100°C in alkaline or neutral medium. Under such condition there is increase in acidity in one case and alkalinity in another. The non-protein nitrogenous compounds are positively correlated with the turbidity in *gur*. Mohan Rao *et al.* (1951) observed a depression in *gur* quality accompanied by rise in inorganic non-sugar and non-protein nitrogen with drop in phosphate.

Polyphenols

Tannins (oxidised polyphenols) derived from protocatechuic acid, phenolic hydroxyl, anthocyanin and saccharetin, are present in cane fibre. Melanoidin, another coloured complex, is formed by the reaction of amino acids with the reducing sugars present in juice, when subjected to high temperature and change in pH. Polyphenols, when present in large quantity, form dark coloured compounds after combining with traces of iron. The enzyme polyphenol oxidase acts on polyphenols in the presence of iron, which is perhaps a co-factor forming various complexes. The phenolic acid and the oxidised products of polyphenol are coloured substances. The tannins in sugar cane juice form a blue green pigment (Zerban, 1918, 1919). Faber and Carpenter (1971) have identified 21 phenols occurring in sugarcane plant, ten of these are carried over to the stage of raw sugar. No positive relationship between tannin and colour of *gur* was found by Vishwanathan (1937-1938). Jatkar and Dangre (1953) while reviewing the effect of iron on colour reported that the colour of *gur* is mostly dependent on iron content rather than its caramelisation.

Organic Acids

Organic acids, such as, malic, succinic, glutamic, glycolic, citric and aconitic are present in juice. The acidity of juice is mainly due to aconitic acid which is a main competitor for phosphoric acid (Narsimham, 1957).

Inorganic Non-sugars

In north India, removal of reversible colloids in juice appears to be a prerequisite for the preparation of better quality of *gur*. Preparation of *gur* in north India suggests the need for using a proteinacious or mucilaginous plant or seed extract to ensure proper crystallisation of sucrose in the manufacture of *gur* (Narsimham, 1957).

Colloidal fraction of total ash is positively correlated with colour, acidity and negatively with pore space (Khanna and Chacravarti, 1955). This shows that higher per cent of colloidal fraction in ash produces dark colour. High acidity and mineral nutrients make the *gur* a porous structure. Moreover, large quantities of ash make the *gur* saltish. Ghosh and Agrawal (1983) have suggested that the upper limit of ash should not exceed six per cent in *gur*.

The following points are also worth mentioning:

(i) the colour of *gur* samples is positively correlated with certain undesirable constituents, such as, reducing sugars, ash content and calcium (Ghosh and Agrawal, 1983).

(ii) magnesium, sesqui oxides (F_2O_3 + Al_2O_3) and calcium oxide content of juice have been correlated with properties of *gur* in the same manner as non-protein nitrogen (Khanna and Chacravarti, 1955)

(iii) most of the iron present in the juice is in complex form and very little is water soluble. Addition of sodium and potassium salts of organic acids enhances the solubility of iron (Farooque and Lala, 1957)

(iv) presence of silicic acid in juice with negative charge reacts with other non-sugars. This helps in the sedimentation and filtration, effecting the quality of *gur*. The ratio of salicic acid and sesquioxide is the main constituent to predict the behaviour of juice in clarification (Honig, 1953).

Alkali Metals (Na, K, Ca and Mg)

The potassium content of juice is 50-70 per cent of total cations present. It is possible to determine the changes in the composition of ash by removal or increase of non-sugars. Farooque and Lala (1957) reported that *gur* colloids are capable of exchanging cations present in the juice. It appears that added calcium ions are absorbed by colloids (Singh *et al.*, 1978). The presence of sodium in exchange complex is associated with higher degree of hygroscopicity of *gur* compared to other ions (K, Mg and Ca). Induced hygroscopicity by these ions is in descending order Na > K > Mg > Ca because smaller the atomic number and size of the absorbed ions, the greater is the number of water molecules in the system. Calcium improves the keeping quality, while sodium affects it adversely. The nature of the exchangeable cations on the colloid complex greatly modifies the moisture holding capacity and dehydration of *gur.* Lower levels of all these inorganic constituents promote superior clarifying properties of end product (Honig, 1953).

Other non-sugar impurities include fibre particles (cellulose, pentosans, gums, lignins, pectins, etc.) which increase the viscosity of juice. Waxes impart turbidity to juice and fats release fatty acids during clarification.

Colour Substances Formed During the Processing of Juice

During processing of juice high temperature, varying pH, air and iron (pan and crusher) induce breakdown of sucrose. The process is known as caramelisation and the products formed are called caramel—a mixture of three anhydrides of sucrose, namely, caramel, caramelene and carameline. Breakdown of sucrose in the alkaline solution produces acids which are brown coloured substances. The organic acids produced from hexoses are acetic, formic, lactic, saccharetic, dihydroxy butyric and their lactones (Zarban, 1947). Besides these other compounds are formaldehyde, acetyl, biacetyl and gluconic acid. Some dark humin like substances are also formed.

Post-Harvest Preservation of Milled Juice for *Gur* Manufacture

Alexander (1973) reported that when juice is treated with sodium silicate (@20-80 μ mol/ml) it can be preserved up to 48 h. Polycide (dichlorophen in soidum hydroxide base)—2 ml/l, a wide spectrum bactericide—fungicide, and Bactrinol (2 bromo-2 nitro propane - 1-3-diol)-500 ppm were found effective in arresting the growth of *Leuconostoc* sp. (bacterium) and preserve the juice up to 48 h. Addition of these biocides did not produce any detrimental effects on *gur* quality. There was little or no residual effect of the Polycides (less than 1 ppm). However, juice treated with Polycide gave characteristic phenolic odour, whereas in case of Bactrinol such an odour was absent. Juice preserved with high concentration of Bactrinol (2000 ppm) produced degradation products which caused inversion of sucrose resulting in poor *gur* quality. *Gur* from Polycide preserved juice was comparatively poor as it absorbed more moisture than the Bactrinol treated juice. Biocides have also been used to minimize the loss in juice quality of harvested cane up to 5 days when cut ends were dipped immediately in Polycide (2 ml/l) or Bactrinol-100 (1000 ppm) or spraying the same on the cane (Chiranjivi Rao *et al.*, 1982; Chiranjivi Rao and Thangavelu, 1987).

Calculation of Yield of *Gur* from Sugarcane Varieties

At the Regional Research Station, Anakapalle (A.P.), studies were conducted to evaluate sugarcane varieties for *gur* making without actually boiling the juice [Anon., (Anakapalle, India), 1989].

The CCS percentage was estimated by the formula: 0.75 [sucrose% - 0.4% (Brix % - sucrose %)] is accounting only for about 45 per cent of variation in *gur* recovery. So developing of an appropriate formula for estimating *gur* recovery is desirable.

From the data recorded from 100 experimental boiling using 40 varieties, relationship of sucrose and non-sugars to *gur* was worked out and is given in Table 4.3 was found to afford determination of the *gur* sucrose of a variety on cane weight reasonably and accurately without actually conducting the juice boilings.

Average difference between the estimated and actual recovery of *gur* on cane weight was observed to be only 3.0 per cent.

Table 4.3: Yield of Sucrose and Non-sugars as related to *Gur*

Particular		Range	Mean
Sucrose yield % on cane weight (X_1)		8.35-13.65	11.10
Non-sucrose yield (%) on cane weight (X_2)		0.59-2.63	1.27
Gur recovery (%) on cane weight (X_3)		9.05-13.45	11.42
Factor	**Correlation coefficient**	**Partial regression coefficient**	***t* observed**
Sucrose yield (%) on cane weight (X_1) and *gur* recovery (%) on cane weight (Y)	+0.844	b Y 1.2 = +0.8235	17.79
Non-sugar yield (%) on cane weight (X_2) and *gur* recovery (%) on cane weight (Y)	–0.286	b Y 2.1 = +0.6972	5.75
t at p = 0.05 ± 0.195			
t at p = 0.05 ± 0.198			

Source: Prasada Rao, 1977.

The equation $Y = 1.3899 + 0.8235\ X_1 + 0.6972\ X_2$

where Y is % *gur* recovery on cane weight
X_1 is % sucrose yield on cane weight
X_2 is % non-sugar yield on cane weight

REFERENCES

Alexander, A.B. (1973). *Sugarcane Physiology*, Elsevier Scientific Publishing Company, Armsterdam, pp. 573-595.

Anonymous (1946). Annual Report of work of Agricultural Chemist. *Proc. Bien. Conf. S.R.D.W.I.* 2: 456.

Anonymous (Padegaon, India). (1933-37). *Annl. Rept. Sugarcane Research Station*, Padegaon, Maharashtra.

Anonymous (Shahjahanpur, India) (1965). *Annl. Rept. Sugarcane Research Work Done in U.P.*, Shahjahanpur, U.P.

Anonymous (Shahjahanpur, India) (1983). *Annl. Rept. Sugarcane Research Work Done in U.P.*, Shahjahanpur, U.P.

Anonymous (Anakapalle, India) (1989). *75 Years of Agricultural Research at Anakapalle (1913-1988)*, Regional Agricultural Research Station, Anakapalle, A.P.A.U., Hyderabad, pp. 400.

Barnes, J.H. (1918). Sugar and the sugarcane in the Gurdaspur District. *Agriculture Research Institute Bulletin No* 69, pp. 100, Supdt. Govt. Printing, India, Calcutta.

Chacravarti, A.S., K. Prasad and K.L. Khanna (1954). A ready made clarifying agent for use in the open pan boiling of sugarcane juice. *Proc. Bien. Conf. S.R.D.W.I.* 3 (Pt. II): 596-602.

Chiranjivi Rao, K., K. Seshadri, I.Y.L.N. Murthy and L. Selvarajan (1982). Polycide - A potential bactericide for cane juice preservation. *Proc. Annl. Conv. S.T.A.I.*, 46: G 55.

Chiranjivi Rao, K., and S. Thangavelu (1987). Bactrinol - 100—a potential and promising biocide for sugar industry. *Proc. Annl. Conv. S.T.A.I.*, 50: G 27.

Chockalingam, S. (1985). *Gur* manufacture in Tamil Nadu-A Review. *National seminar-cum-group discussion on jaggery manufacture and storage,* held at IISR, Lucknow (Dec. 18-19).

Faber, L. and F.G. Carpenter (1971). Identification of cane pigment that persists into refined sugar, *Proc. ISSCT* 14: 1586-1607.

Farooque, M. and S.C. Lala (1957). Studies in chemistry of gur in relation to clarification of partially limed juice *Indian J. Sugarcane Res. and Devpt.* 1 (3): 139-142.

Ghosh, A.K. and M.P. Agrawal (1983). Gur grading based on physical and chemical constituents. *Maharashtra Sugar* 8(12): 39-43.

Gopala Aiyer, K.V. and K.C. Rao (1954). Juice characteristics of sugarcane varieties and their relation to quality of jaggery. *Proc. Bien. Conf. S.R.D.W.I.* 2 (II): 564-571.

Hadi, S.M. (1929). The Indian Sugar Industry, Govt. of Bhopal, C.I., Published by Khan Bahadur S.M. Hadi, Director of Agriculture, Bhopal State, printed by F.W. Petch Thacker and Co., Ltd., Bombay, pp. 289 + vi.

Honig, P. (1953). *Principles of Sugar Technology,* Elsevier Publishing Co., Amsterdam.

Jatkar, S.K.K. and A.J. Dangre (1953). Role of iron in the colouration of *gur*. *Proc. Annl. Conv. D.S.T.A.* 10: 56-61.

Khanna, K.L. and A.S. Chacravarti (1955). Scientific Monograph—Researches on Technical Aspects Relating to Improvement of Gur Industry in Bihar, Indian Central Sugarcane Committee, New Delhi.

Krishnamurty, K. (1919). The effect of salinity on the growth and composition of sugarcane. *Agric J. India* 14: 476-493.

Leley, V.K., S.B. Kaderkar and R.A. Kale (1955). Improving the colour of *gur*. *Annl. Conv. D.S.T.A.* 12 (Pt. I): 56-61.

Mohan Rao, N.V. and D. Anji Reddy (1951). Chemical composition and keeping quality of jaggery from cane grown under swamp conditions. *Proc. Bien. Conf. S.R.D.W.I.* 1(2): 152-156 (1951).

Mohan Rao, N.V., R.L. Narsimham and D. Anji Reddy (1951). Studies on quality of Jaggery. *Proc. Bien. Conf. S.R.D.W.I.* 1 (2): 143-151.

Narsimham, R.L. (1957). Regional variation in the composition of sugarcane juice in relation to claribility. *Proc. Annl. Conv. S.T.A.I.* 25: 27-42.

Panje, R.R. and A. Singh (1963). Effect of iron crusher and boiling pan on colour of *gur* of different cane varieties. *Indian J. Sugarcane Res. Devpt.* 7(3): 197-198.

Prasada Rao, K.K. (1977). Note on calculation of jaggery from sugarcane varieties. *Indian J. agric. Sci.* 47(1): 58-59.

Satyanarayana, K.V.S. (1939-43). *Report on the Scheme for Research on Chemistry of Sugarcane,* Indian Agricultural Research Institute, New Delhi.

Singh, M., K.M. Bharadwaj and M.L. Agarwal (1978). Storage of *gur*—A Review.

Co-op. Sugar 9(11): 559.

Vishwanathan (1937-38). *Report on the Scheme for Research on Chemistry of Sugarcane*, Indian Agricultural Research Institute, New Delhi.

Wiggins, L.F. and J.W. Williams (1951). *Proc. Meeting British West Indies Sugar Technol.*: 40-45.

Zerban, F.W. (1918). *Industrial Engineering Chemistry*, vol. 10: 1034.

Zerban, F.W. (1919). *Industrial Engineering Chemistry*, vol. 11: 814.

Zerban, F.W. (1947). *Sugar Res. Foundation Technol. Rept.*, Series No. 2.

5

INFLUENCE OF CULTURAL PRACTICES (FOR RAISING SUGARCANE CROP) ON *GUR* YIELD AND QUALITY

Cultural practices followed in raising sugarcane crop appreciably affect growth and development of the crop, pest and disease build up and response to stress, particularly drought as also the sugar content. In the earlier part of this century, in India, experiments on the influence of manuring and spacing on *gur* yield were conducted at Sabour, in Bihar (Taylor, 1913) and Pratabgarh (Clarke *et al.* 1912). Afterwards, in Uttar Pradesh, with the inception of the scheme for the development of sugarcane in *gur* and *khandsari* areas, Division of Agronomy (*Gur*) was established at the Sugarcane Research Station, Shahjahanpur in 1956-57. Studies conducted on the influence of cultural practices followed for sugarcane on *gur* yield and quality are mentioned below:

(a) *System of Planting:* Experiments conducted at Raya (Mathura), Kalai (Aligarh) and Varanasi from 1960 to 1967, showed that double row planting significantly improved sugarcane yield over single row planting; however, the yield and quality of *gur* did not differ significantly (Anon., (Shahjahanpur, India), 1983).

(b) *Time of Planting:* In subtropical India, planting cane towards the end of January, end of February and end of September were best from the point of view of *gur* yield. April planting proved worst giving an average of 3.339 mt of *gur* per ha which was 53 per cent of that obtained from February planted sugarcane. In quality, the *gur* from March and April planted cane proved to be the worst. Best quality of *gur* was obtained from October, November and January plantings. September planted crop gave good quality *gur* during early season but did not prove of advantage over October, November and January plantings during the later part of the crushing season (Misra, 1967).

Table 5.1: Effect of Planting Date on Yield (t/ha) and Quality of *Gur*

Attribute	Date of planting cane							
	Sept.	Oct.	Nov.	Dec.	Jan.	Feb.	Mar.	Apr.
Gur yield (t/ha)	6.18	5.09	5.58	4.92	6.24	6.49	4.89	3.36
Pol per cent *gur*	77.6	79.2	79.0	78.3	79.2	78.8	77.2	73.6
Purity coefficient	83.1	85.6	85.2	84.5	85.2	85.0	83.2	79.1
Invert sugar % *gur*	7.6	6.6	6.2	6.9	6.2	6.8	7.5	10.3
Ash % *gur*	2.15	1.88	1.96	2.13	1.96	2.09	2.03	2.00

Source: Misra, 1967.

Generally, in subtropical India, 12 months spring planted crop and 14 to 16 months autumn planted crop (early in the season) are used for *gur* making. In Maharashtra, parts of Andhra Pradesh and Karnataka, 18 to 20 months crop called *adsali* is also used for *gur* making besides *eksali* crop (Patil and Shingte, 1982).

(c) *Application of Pesticides at Planting:* Studies conducted at Shahjahanpur indicated that application of heptachlor (5.6 l (20% EC) in 1685 l of water ha^{-1}) + aretan (6% Hg @ 280 g in 112 l of water ha^{-1}) increased the yield of cane and *gur* per ha over control by 21.4 t and 2.74 t ha^{-1}, respectively (Misra and Singh, 1964)

(d) *Earthing and Propping:* It has been observed that lodging not only affects the yield of *gur* and its quality (Rege and Tembhekar, 1954) but also it quickly deteriorates during storage (Lal *et al.*, 1985). Earthing and propping to prevent lodging improved recovery of *gur* on per cent juice and cane weight basis in subtropical (Anon., (Shahjahanpur, India), 1983) and tropical India (Chockalingam, 1985).

(e) *Spacing:* In the early part of this century, Clarke *et al.* (1912) studied the effect of spacing on cane and *gur* (the raw sugar) yield at Pratabgarh (in erstwhile United Provinces) and indicated that 29652 to 37065 setts ha^{-1} were sufficient to ensure maximum cane yield. The yield of *gur* did not increase beyond 29652 setts ha^{-1}. At Kalai (Aligarh), during 1966-69, double row planting with 3-bud setts (@ 1 sett per 30.5 cm row length) gave the highest yield of 67.25 t cane and 7.13 t *gur* ha^{-1}, respectively, whereas an inter-row spacing of 60 cm with one sett per 45 cm row length, gave the lowest cane yield. However, the latter gave relatively better quality *gur* as compared to other treatments (Anon., (Shahjahanpur, India), 1983). Work done at Pusa (Bihar) indicated that wide spacing affected *gur* colour and it could be improved by irrigations (Srivastava and Prasad, 1954b).

(f) *Time of Harvest:* In the earlier part of this century, Barnes (1918) emphasised that for production of a dry and light coloured *gur,* the ripeness of cane at the time of harvest is important. Studies conducted at Shahjahanpur from 1958-59 to 1961-62 indicated that *gur* recovery (as per cent of juice) improved gradually with the advancement of season from October to April with a slight depression in the relatively colder part of the season. With advancement in the season and increase in pol per cent juice, purity was associated with decrease in reducing sugars, ash per cent, moisture content and colour of the *gur* prepared till February and March. For the storage purpose, the *gur* prepared during January to March was superior as compared to *gur* produced during early or late in the season (Anon., (Shahjahanpur, India), 1983). In the tropical India also, quality *jaggery* could be obtained from the crop harvested from January to March. Harvesting premature (Rege and Tembhekar, 1954) and over mature (Chockalingam, 1985) crop is neither conducive for the production of *gur* nor its quality. Immature crop contains less sucrose and more of reducing sugars and non-sugars, whereas in over mature crop inversion takes place (Chockalingam, 1985). At Kolhapur, in Maharashtra, quality *gur* was obtained from a mid-late ripening variety Co 7527 at the age of 14 months. Harvesting the crop earlier to it did not yield good quality *gur* (Wandre and Hasabnis, 1991).

(g) *Gap between Harvest, Crushing and Boiling of Juice to Gur:* In the earlier times, amongst the farmers of Gurdaspur district

(Punjab) it was a common practice to allow one day lapse between harvesting and crushing as it yielded relatively drier and better quality *gur* (Barnes, 1918).

For making *gur* from the red rot affected canes, the juice extracted should be boiled immediately after the crushing otherwise it may not set well (Hadi, 1929).

(h) *Barani Conditions:* In the beginning of this century, it was observed that the quality of *gur* from a *barani* crop of sugarcane was superior to the one obtained from an irrigated crop (Barnes, 1918). In the low lying areas or the river banks and *sailab* (flooded), the crop never ripens properly and *rab* is made instead of *gur* (Barnes, 1918). Studies conducted at Shahjahanpur indicated that under rainfed conditions, planting sugarcane in 60 or 90 : 30 : 90 cm apart rows and the application of 75 kg N ha^{-1} through basal or 50 kg N ha^{-1} through basal and 25 kg N as foliar spray, resulted in more yield of quality *gur* than planting at the same spacing and applying 75 kg N ha^{-1} in other ways (Anon., (Shahjahanpur, India), 1983).

(i) *Intercropping:* Studies conducted at Shahjahanpur and Raya (Mathura) during 1970-71 and 1972-73 indicated that autumn planted sugarcane intercropped with potato, wheat and *lahi* gave markedly higher yield of cane and *gur* than the spring planted crop taken after *rabi* crops. Studies conducted later at Kalai (Aligarh) also confirmed these results. Besides, it was observed that autumn planted cane gave the highest yield of sugarcane and *gur* as in comparison to one intercropped with *rabi* crops. At Shahjahanpur, intercropping peas with autumn cane yielded better quality *gur* as compared to other intercrops like *Bengal gram, berseem* and *lentil* (Anon., (Shahjahanpur, India), 1983). Sugarcane intercropped with *lobia* (for fodder) supplied with 100 or 10 kg N ha^{-1} yielded superior quality of *gur* (Anon., (Shahjahanpur, India) 1982-83.

(j) *Manuring:* As early as in 1869, in Java (now Indonesia), Krajenbrink observed that too heavy nitrogenous manuring produced a weak juice (*Sugar Cane* Vol. I, pp, 37; 93). In India, Leather (*Agricultural Ledger, 1897,* No. 3) at Cawnpur (= Kanpur), observed weakening of juice of *Matna* variety of sugarcane by application of nitrogenous manure (*c.f.* Taylor, 1913). Experiments by Taylor (1913), at Sabour, indicated that the stage of peak ripening in the crop top dressed with nitrogenous manure is achieved a fortnight earlier than the crop which was not top dressed with nitrogen.

At Shahjahanpur, Misra *et al.* (1965) observed that the differences in either yield or quality of *gur* due to application of fertilizers were not consistent. Amongst nitrogenous fertilizers, ammonium chloride affected the *gur* quality adversely by increasing reducing sugars and ash contents thereby decreasing net rendements.

Further studies indicated that recovery and quality of *gur* were slightly better in the unmanured crops as compared to the crops where the nitrogen was applied in two splits. The quality of *gur* as well as its keeping quality both are adversely affected due to increasing levels of nitrogen manuring. The *gur* made from a plant crop sugarcane of variety Co 419 supplied with higher nitrogen (224 kg N ha^{-1}) was relatively more acidic, contained more nitrogen, moisture, reducing sugars and was dark in colour. It contained less ash, phosphorus, sucrose, and lesser pore space as compared to the one obtained from a crop raised with 112 kg N ha^{-1} (Anon., 1989). Delayed application of nitrogen results in inferior quality of *gur* (Chockalingam, 1985).

Interaction of nitrogenous fertilizer and irrigation also influenced the recovery of *gur.* The most suitable interval for irrigation during pre-monsoon period was 15 days. Increasing the dose of nitrogen raised the recovery of *gur* per cent juice under frequent irrigations, while reserve was observed when irrigation was given at 21-day intervals (Anon., 1983).

More quantity of *gur* is obtained from a sugarcane crop supplied with 150 kg N ha^{-1} in 1 : 2 or 1 : 1 ratio through organic manure and inorganic fertilizer. Relatively better quality of *gur* is obtained from a green manured sugarcane crop (Anon., 1983).

There is another aspect of *gur* making which is influenced by manuring. Nitrogenous manuring tends to adversely affect juice claribility as also the colour of *gur,* while effect of P is highly beneficial. The injurious effects of higher applications of nitrogen could be minimised if applied in combination with

P and four to six pre-monsoon irrigations (Srivastava and Prasad, 1954, 1954a, 1954b). At Malalathur, in Tamil Nadu, application of 75 kg P_2O_5 and 112 kg K_2O ha^{-1} improved yield and quality of *gur* (Chockalingam, 1985). At Kolhapur, in Maharashtra, application of up to 200 kg K ha^{-1} and harvesting the crop of mid-late ripening variety Co 7527 at the age of 14 months improved *gur* quality as well as its storability (Wandre and Hasabnis, 1991).

For a detailed discussion on the effect of soils and fertilizers on *gur* quality, Chapter 6 may be consulted.

(k) *Ratooning:* Investigations at Shahjahanpur revealed that the *gur* prepared from a ratoon crop exhibited better keeping quality. An unirrigated ratoon crop with full dose of nitrogen of 67.56 kg ha^{-1}, applied late in the season at the break of the monsoon, resulted not only in lower recovery of *gur* per cent cane and yield of *gur*, but also it deteriorated rapidly (Anon., (Shahjahanpur, India), 1983).

Plant crop of some of the sugarcane varieties like Co T 8201 gave good yield of *gur* (15.17 t ha^{-1}) but its ratoon crop did not. But in some other varieties, like Co 6907 and Co 7706, ratoon crop also proved superior in cane as well as in *gur* yield (Bhupala Rao *et al.*, 1991).

Increasing number of pre-monsoon irrigations from 1 to 3 resulted in marked improvement in yield of cane and *gur* per ha. The application of 150 kg ha^{-1} and 4 pre-monsoon irrigations to the ratoon crop of sugarcane variety CoS 510 gave higher yield of *gur* as compared to 100 kg N ha^{-1} with 3 pre-monsoon irrigations and 200 kg N ha^{-1} + 50 kg P_2O_5 ha^{-1} with 5 pre-monsoon irrigations (Anon., 1983).

REFERENCES

Anonymous (Anakapalle, India) (1989). *75 Years of Agricultural Research at Anakapalle (1913-1988)*, RARS, Anakapalle, A.P.A.U., Anakapalle, pp. 400.

Anonymous (Shahjahanpur, India) (1983). *70 Years of Sugarcane Research (1912-1981)*, U.P. Council of Sugarcane Research, Shahjahanpur, U.P.

Anonymous (Shahjahanpur, India) (1982-83). Annl. Rept. Sugarcane Research in Uttar Pradesh, U.P. Council of Sugarcane Research, Shahjahanpur, U.P.

Barnes, J.H. (1918). *Sugar and the Sugarcane in the Gurdaspur District*, Agricultural Research Institute, Pusa (Bihar), *Bulletin No. 69*, pp. 100, Supdt. Govt. Printing, India, Calcutta.

Bhupala Rao, J.V.R., V. Raja Bapa Rao, T. Anuradha and S. Ramakrishna Rao (1991). Sugarcane varietal performance in plant and ratoon crops in relation to jaggery quality, presented at the *National Seminar on Increasing Productivity, Quality and Storability of Jaggery (Gur), (Nov., 30-Dec., 1)*, organised by Indian Institute of Sugarcane Research, Lucknow and Indian Society of Agricultural Engineers, New Delhi, at the Indian Institute of Sugarcane Research, Lucknow, India.

Clarke, G., H.E. Annett, S.Z. Husain, S.C. Banerji and N. Hussain (1912). *Experiments on the Cultivation of Sugarcane at the Pratabgarh Experiment Station: 1909-1911.*, Agricultural Research Institute, Pusa (Bihar), pp. 29, Supdt. Govt. Printing, India, Calcutta.

Chockalingam, S. (1985). Gur manufacture in Tamil Nadu-A review. *Proc. National Seminar-cum-Group Discussion on Jaggery (Gur) Manufacture and Storage (Dec., 18-19, 1985)*, Indian Institute of Sugarcane Research, Lucknow, India, pp. 47-50.

Hadi, S.M. (1929). The Indian Sugar Industry, Govt. of Bhopal, C.I., printed by F.W. Petch Thacker and Co. Ltd., Bombay, published by Khan Bahadur S.M. Hadi, Director Agriculture, Bhopal State, pp. 289 + iv.

Hasabnis, A.B., S.S. Wandre and A.G. Shinde (1991). Screening of sugarcane varieties harvested at different stages for jaggery quality, presented at the *National Seminar on Increasing Productivity, Quality and Storability of Jaggery (Gur), (Nov., 30-Dec., 1)*, organised by Indian Institute of Sugarcane Research, Lucknow and Indian Society of Agricultural Engineers, New Delhi, at the Indian Institute of Sugarcane Research, Lucknow, India.

Lal, U., S.C. Srivastava, R.K. Sharma and S.K. Verma (1985). Studies on improving the quality of *gur* prepared from lodged cane. *Proc. National Seminar-cum-Group Discussion on Jaggery (Gur) Manufacture and Storage (Dec., 18-19, 1985)*, Indian Institute of Sugarcane Research, Lucknow, India, pp. 47-50.

Misra, G.N. (1967). Effect of time of planting on yield and quality of *gur*. *Proc. Annl. Conf. S.T.A.I.* 35: 23-30.

Misra, G.N. and J.D. Singh (1964). Improving crop stand and final yield of sugarcane and *gur* through application of chemicals (fungicides and insecticides) at planting time. *Proc. All India Conf. S.R.D.W.I.* 5: 119-131.

Misra, G.N., B.S. Mathur and J.D. Singh (1965). Response of sugarcane to different nitrogenous fertilizers: Effect on yield of cane and *gur* and its quality. *Proc. Annl. Conv. S.T.A.I.* 33: 39-48.

Patil, J.P. and A.K. Shingte (1982). Studies on the effect of different levels of press mud, K and nitrogen on the quality of jaggery. *Indian Sugar* 32 (3): 167.

Rege, R.D. and V.V. Tembhekar (1954). Studies on the influence of brix and purity of juice on the production of *gur*. *Proc. Bien. Conf. S.R.D.W.I.* 2 (II): 572-574.

Srivastava, D.P. and K. Prasad (1954a). Effect of manurial and cultural treatments on the claribility of sugarcane juice in the open pan boiling process. I. Nitrogen-phosphorus manuring and irrigation. *Proc. Bien. Conf. S.R.D.W.I.* 2 (Pt. II): 584-591.

Srivastava, D.P. and K. Prasad (1954b). Effect of manurial and cultural treatments on the claribility of sugarcane juice in the open pan boiling process. II. Nitrogen dose in relation to irrigation and spacing between rows of crop. *Proc. Bien. Conf. S.R.D.W.I.* 2 (Pt. II): 591-596.

Taylor, C.S. (1913). Notes on experiments with sugarcane at Sabour Pt. I On the character and quantity of fibre in different varieties, Pt. II On the effect of different manurial treatments on the time of ripening of cane, Agricultural Research Institute, Pusa (Bihar), *Bulletin No. 37*, pp. 17, Supdt. Govt. Printing, India, Calcutta.

Wandre, S.S. and A.B. Hasabnis (1991). Effect of different potash levels and harvesting dates on quality of jaggery, presented at the *National Seminar on Increasing Productivity, Quality and Storability of Jaggery (Gur), (Nov., 30-Dec., 1),* organised by Indian Institute of Sugarcane Research, Lucknow and Indian Society of Agricultural Engineers, New Delhi, at the Indian Institute of Sugarcane Research, Lucknow, India.

6

EFFECT OF SOIL AND FERTILIZER ON QUALITY OF *GUR*

Old literature (Anon., 1877-78) indicates that in district Pratabgarh in the erstwhile Oudh Province of British India, best quality *gur* having light yellow colour and fetching high price was made in Pargana Patti irrespective of sugarcane variety than at other areas. In the erstwhile Bhopal State, a sugarcane variety *Khajla* producing the best quality *gur*, was practically confined to villages of *Maina* and *Khamkhera* in Ashta *tehsil*. It did not flourish in other cane growing areas of State (Hadi, 1929). Further, poor quality *gur* is obtained from the juice of cane growing in saline soils (Krishnamurthy, 1919). Superior quality of *gur* is produced in the district of Kolhapur in Maharashtra (Ranadive, 1985) and in Azamgarh in Uttar Pradesh (Hoda, 1985). Studies conducted later indicated that the superior quality of *gur* made at Kolhapur was associated with low salt content of the soil supporting sugarcane growth (Ranadive, 1985). These evidences indicate that chemical composition of the soil supporting sugarcane growth has a profound influence on the quality of *gur*.

As early as in 1869, Krajenbrink in Java (now Indonesia) stated "too heavy nitrogen manuring produced a weak juice". In India also, Leather, working at Cawnpur (= Kanpur) observed weakening of juice of *Matna* variety of sugarcane by applying nitrogenous manure (*c.f.* Taylor, 1913). Experiments conducted during the early part of this century at Sabour, in Bihar, indicated that application of nitrogenous manure delayed the ripening by a fortnight (Taylor, 1913). Beside, it impaired the quality of *gur* at Gurdaspur, in Punjab (Barnes, 1918). These call for a detailed discussion on the influence of soil characteristics and fertilizers on the quality of *gur*. This aspect has been discussed in detail in this chapter.

Effect of Soil Type on Quality of *Gur*

Basically sugarcane juice is a plant (stem) sap which contains abundant quantity of sucrose and little amounts of glucose and fructose. It also contains some non-sugars (nitrogenous material, organic substances and minerals). These compounds are formed largely by different metabolic processes. Such metabolic activities depend on environment, soil type and cultural practices during cane cultivation. In India sugarcane is cultivated in five different agroclimatic zones (Fig. 6.1) and soil types. These profoundly affect the proportion of sugars to non-sugars in juice. As a result, clarification of juice requires certain treatments for removal of these non-sugars.

Van Dillewijn (1948) stated that relatively more pure juice was obtained from cane growing in soils containing lime stone and higher salt content.

Juice from cane varieties growing on alluvial soil containing alkalies, contained less sugar and produced more molasses as compared to normal soils (Deer, 1931). In Mauritius, alluvial soils favour luxuriant vegetative growth in wet season but the cane stalks remain watery. Conversely, calcareous soils produce relatively thicker canes than the clayey soils containing organic matter. Poor quality cane was produced in different types of poorly drained soils in Puerto Rico (Lugo Lapez, 1954).

Composition of Cane Juice vis-a-vis *Soil Type*

UTTAR PRADESH AND PUNJAB

When same varieties are cultivated in Uttar Pradesh and Punjab, the mineral content is higher in arid tracts but the content of soluble salts do not differ. Although both States have alluvial soils, the recovery of *gur* is relatively more in Uttar Pradesh as compared to Punjab (Lender and Narain, 1936).

COASTAL ANDHRA PRADESH AND ORISSA

In coastal Andhra Pradesh and Orissa where soils are loamy (or alkaline) the juice is acidic (pH: 5.2-5.4) in nature. In parts of Orissa where soils are lateritic, the acidic juice contains low

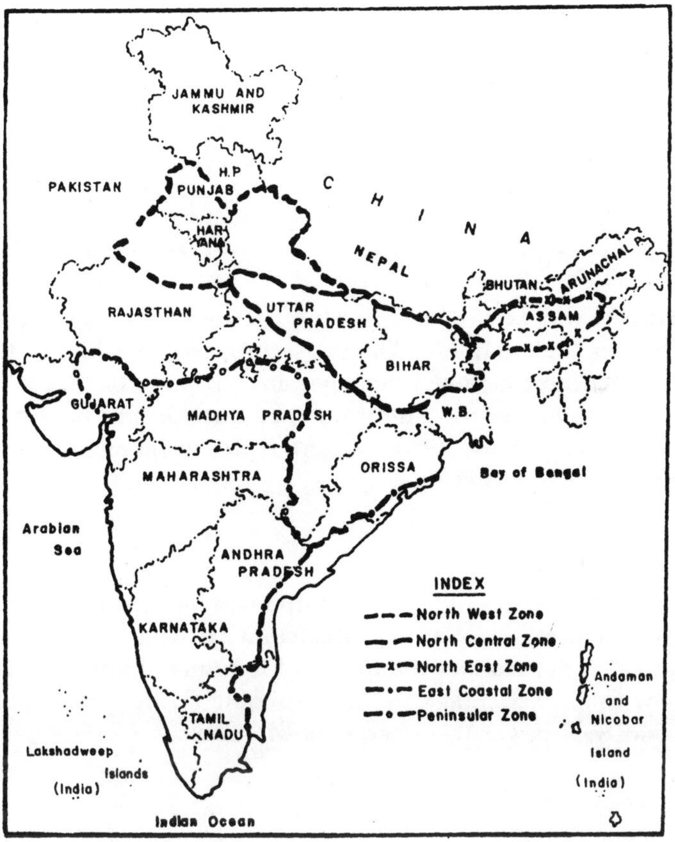

Fig. 6.1: Sugarcane Cultivated in Different Agroclimatic Zones in India

divalent cations and has high silica content which does not allow proper setting of *gur* even with high liming (of juice). However, in parts of Chittoor tract of Andhra Pradesh having comparatively high soil pH, normal setting of *gur* is attained by simply boiling the juice obtained from well ripe canes.

KERALA

The lateritic soils of Kerala have low pH (5.2) and low divalent cations but are rich in soluble silica. When *gur* is manufactured from cane crop grown on such soils, it does not set properly.

KARNATAKA AND MUMBAI (DECCAN TRACT, MAHARASHTRA)

At Hospet (Karnataka) and in Mumbai Deccan tract, the soil pH is very high (pH 9.5). The *gur* setting can be achieved by simply boiling the juice from ripe canes. In the same tract, in soils having higher content of lime, addition of phosphate in juice (or phosphatic fertilization) is essential to improve claribility of the juice.

BIHAR

Available P_2O_5 status of the soil is important for superior quality *gur,* whereas excess quantity of calcium carbonate in soils of north Bihar deteriorates the quality. Medium to light textured soils with good drainage are most suitable for the production of good quality *gur* (Mukherjee *et al.,* 1988).

MAHARASHTRA

Crop grown in deep black soils of Maharashtra which are rich in nutrients generally produce dark coloured *gur.*

(i) *Kolhapur*: Soils are silty loam having fairly less clay complex and base exchange capacity which may account for relatively lesser accumulation of nutrients in the juice. This results in higher purity and recovery with light coloured *gur.*
(ii) *Padegaon:* It represents typically arid climate. Soils of this area have expanding montmorillonite clay mineral,

more salt accumulation and possess high base exchange capacity. The cane produced contains higher amount of nutrients, particularly, nitrogenous compounds which impart dark colour to *gur* under normal conditions of manufacture. Large quantity of clarificants is added to the juice to obtain a golden yellow colour of the *gur.* In the absence of adequate rainfall the crop needs to be irrigated. If more irrigations are given, there is less sugar adversely affecting the recovery of *gur* (Dagade and Ranadive, 1988).

Effect of Soil Factors

SALT AFFECTED SOILS

There is a depression in juice purity with concomitant increase in potash and chlorine content in juice when the crop is grown on saline soil. *Gur* prepared from such canes is saltish and of poor keeping quality (Krishnamurthy Rao, 1919; Swadi, 1919; Lender and Narain, 1936). The immediate action of salts appears to be due to a reduction in the precipitation of colloids when clarificants are added.

WATERLOGGED SOILS

Flooding of a crop from July to September with 120 cm deep water increased the content of non-protein nitrogen (90% of the N present) and decreased the sucrose and gum content in the cane juice (Khanna and Chacravarti, 1949). Under such conditions, the contents of glucose and ash increased and that of P_2O_5 decreased (Anon., (Ankapalle, India), 1989). Such an imbalance results in poor juice quality and loss in *gur* recovery. In the low lying lands along the river or canal embankments or on the banks of swamps of Punjab where there is no drainage, no variety other than *Katha* grew. Under such conditions, *Rab* is generally made instead of *gur* as the crop never ripens properly (Barnes, 1918).

The *gur* prepared from cane growing in swampy conditions contained relatively higher total as well as non-protein nitrogen that the cane growing in normal conditions (Table 6.1). Besides, it absorbed relatively more moisture at 75 per cent relative humidity (RH). These could be the possible causes for relatively quicker deterioration of the *gur* prepared under such conditions (Anon., (Anakapalle, India), 1989).

Table 6.1: Composition of *Gur* from Cane Growing in Swampy Conditions

Particular	Condition	
	Swampy	Normal
Moisture (%)	7.09	6.97
Sucrose (%)	74.86	88.37
Purity coefficient	74.86	88.37
Glucose (%)	9.59	6.55
Glucose ratio	12.81	7.41
Total non-sugars (%)	15.55	5.08
Nitrogen (mg /100 g *gur*)		
(a) Total	95.00	76.40
(b) Non-protein N	67.27	50.95
(c) Protein N	27.73	25.45
Non-protein N (as per cent of total)	70.82	66.70
Colloids (%)	5.14	4.91
Moisture absorbed at 75% R H	5.35	1.75

Source: Anon., (Anakapalle, India), 1989, p. 254.

Effect of Fertilizer on *Gur* Quality

In India, as early as in 1918, it was observed that over manuring of canes decreased quality of *gur* produced (Barnes, 1918). Studies conducted later on led to following fertilizer recommendations for producing quality *gur* in India (Table 6.2).

Nitrogenous Fertilizers

In India, sugarcane crop is fertilized by inorganic nitrogenous fertilizers, namely, urea, ammonium sulphate with or without organic manures. In 1950s, ammonium sulphate was the choice of the farmers as it boosted sugarcane production per unit area but it was subsequently withdrawn as it adversely affected the juice quality. Thereafter, urea became the favourite of all the farmers as it appreciably increased yield and also improved the juice quality. The nitrogen manuring tends to affect juice claribility and also the colour of the end product (Prasad and Srivastava, 1954). Asokan *et al.* (1978) reported that urea was at par with ammonium sulphate with regard to quality of *gur*. Although higher levels of N are conducive to higher production of cane these, at the same time adversely affect the juice quality by increasing glucose and concomitantly decreasing the sucrose content (Anon., (Shahjahanpur, India), 1983; Rege and Tembhekar, 1954; Misra *et al.*, 1965; Wandre *et al.*, 1985). *Gur* prepared from such a crop containing higher amounts of glucose, a hygroscopic aldose monosaccharide, absorbs atmospheric moisture and, therefore, the deterioration of *gur* is relatively faster when stored after July (Javelekar, 1979).

Table 6.2: Recommended Levels of Fertiliser for Producing Quality *Gur*

State	Location	Fertiliser (kg ha^{-1})			Reference
		N	P_2O_5	K_2O	
A. **Subtropical India**					
(a) Uttar Pradesh,	Lucknow	45	—	—	1
	Shahjahanpur	75	50		2
		150			
B. **Tropical India**					
(a) Maharashtra,	Kolhapur	200	115	115	3
(b) Tamil Nadu,	Cuddalore	225	75	112	4
	Melalathur	175	75	112	5
(c) Andhra Pradesh	Anakapalle	112	—	—	6

1. Agarwal and Ghosh (1983).
2. Lal *et al.* (1991).
3. Wandre *et al.* (1985).
4. Shakunthala (1981).
5. Chockalingam (1985).
6. Anon. (Anakapalle, India) (1989).

In subtropical India, the colour of *gur* from cane fertilized with lower levels of N (45 kg ha^{-1}) was relatively better than that from canes fertilized with higher levels of N (113 kg ha^{-1}) (Agarwal and Ghosh, 1983). Quality of *gur* is adversely affected with increasing levels of N (Anon., (Shahjahanpur, India), 1983; Misra *et al.*, 1965). Delayed application of N results in inferior quality of *gur* (Tembhekar, 1954). According to Wandre *et al.* (1985) application of 200 kg N ha^{-1} gave the highest sucrose and purity per cent in juice (20.4 and 91.1 per cent, respectively) and produced higher quantity of quality *gur* as compared to higher doses of 250 kg N ha^{-1}. They further claimed that application of 200 kg N ha^{-1} with 115 kg ha^{-1} each of phosphorus and potash was beneficial with regard to quality and storability of *gur.* Quantity of *gur* was appreciably affected by varying doses of N (Table 6.3). The increase in the amount of N dose generally increases the soluble N in juice which in turn reacts with glucose at the time of boiling giving reddish black colour of *gur.*

Table 6.3: Effect of N Levels on Quality of *Gur* (var. Co 740) in the Month of August

N level (kg ha^{-1})	Moisture (%)	Reducing sugars (%)	Non-reducing sugars (%)	Colour intensity
150	11.2	10.9	67.5	120
200	13.5	14.7	66.5	135
250	14.9	15.5	60.8	150

Source: Wandre *et al.*, 1985.

In another study, also at Kolhapur (Maharashtra), Dagade and Ranadive (1980) reported that increasing levels of N application increased the N, P, K, Ca and Mg content of *gur.* At N level of 375 kg ha^{-1}, the *gur* obtained was of inferior quality and it was of dark in colour. Increasing levels of N also increased the pH of the *gur,* this adversely affected the keeping quality of *gur* with its running off starting at 70 per cent relative humidity (Table 6.4).

Table 6.4: Effect of N Fertilisation on Mineral Content of *Gur* (Cv. Co 775)

Treatment (N kg ha^{-1})	Per cent (on dry weight basis)						
	Moisture	Ash	Nitrogen	Phos-phoric acid	Potash	Calcium	Magnesium
0	5.48	1.42	0.04	0.11	18.10	0.95	0.75
125	6.30	1.52	0.06	0.14	18.28	1.08	0.76
250	6.23	1.53	0.06	0.15	18.66	1.10	0.78
375	6.43	1.54	0.06	0.17	18.90	1.13	0.80
Control	5.68	1.49	0.07	0.21	22.70	1.80	0.86

Source: Dagade and Ranadive, 1980.

Sources of Inorganic Nitrogen Fertilization

Considerable amount of work has been done on the sources of nitrogen *vis-a-vis gur* quality. Misra *et al.* (1965) reported that different forms of inorganic nitrogenous fertilizers do not appreciably affect the yield of *gur* quality. However, ammonium chloride affected the quality of *gur* adversely by increasing the invert sugars and ash content thereby decreasing the net rendement more than others (Table 6.5). Urea, on the other hand, showed comparatively better effect on the quality of *gur* by lowering the quality of invert sugars and ash as shown in Table 6.5.

Table 6.5: Effect of Fertilisers on Different Quality Attributes of *Gur* (average of five locations)

Quality attribute	Ammonium sulphate	Ammonium nitrate	Ammonium chloride	Urea	Control
Pol % *gur*	78.70	79.20	79.30	79.40	80.59
Purity coefficient	83.30	83.30	83.70	83.80	84.70
Invert sugar (%)	5.16	5.02	5.53	4.97	4.18
Ash (%)	3.06	3.01	3.10	2.96	3.06
Net rendement	63.14	63.88	62.95	64.91	65.59
Colour of *gur*	168	135	147	150	134

Source: Misra *et al.*, 1965.

In Andhra Pradesh, application of more than 90 kg N ha^{-1} depressed the quantity and quality of *gur* when irrigation was given to the crop at 75 per cent depletion of available soil moisture (ASM) in the formative phase (Anon., (Anakapalle, India), 1964). Cane juice samples producing quality *gur* have low N/P_2O_5 ratio than those giving a relatively dark coloured *gur* (Tembhekar, 1954). The N/P_2O_5 ratio of 1.0 : 1.5 to 1.0 : 2.5 in the juice is conducive for the production of quality *gur* (Tembhekar, 1954). Applications of low levels of nitrogen and high levels of P_2O_5 in soil, and the presence of low quantity of colloids in juice give good quality *gur*.

Effect of Potassium vis-a-vis *Quality of Gur*

Potassium is essential for improving the health of the crop. Its application also decreases the incidence of various insect-pests and diseases.

The effect of potassium *vis-a-vis gur* quality has been studied in depth by the scientists of the Regional Sugar and *Jaggery* Research Station, Kolhapur (Maharashtra). Patil and Shingte (1982 a) studied the effect of different levels of potassic fertilizers on the quality parameters of *gur* (Table 6.6).

Table 6.6: Effect of Different Levels of Potassic Fertiliser on *Gur* Quality in Different Varieties of Sugarcane

Treatment	Sucrose (%)	Glucose (%)	Moisture (%)	Lead number	Colour intensity (green filter)
Levels of K_2O (kg ha^{-1})					
0	76.10	5.11	6.38	22.76	105
100	84.54	4.05	5.70	20.98	100
200	80.54	4.09	5.53	21.01	101
300	80.55	4.07	5.51	21.05	102
SE +	0.51	0.67	0.81	0.44	0.72
CD at 5%	1.52	NS	NS	1.31	2.13
Sugarcane variety					
Co 740*	77.50	4.70	5.86	22.40	109
Co 1163*	81.36	3.96	5.71	20.46	96
SE ±	0.63	0.018	0.084	0.13	0.89
CD at 5%	2.83	0.08	NS	0.60	4.02

*Sugarcane crop harvested at 12-month stage.
Source: Patil and Shingte, 1982a.

These investigators reported that application of 100 kg K_2O ha^{-1} improved the sucrose per cent, colour intensity and lead number over the control (0 kg K_2O ha^{-1}). The sucrose content of *gur* was higher, while colour intensity, lead number and glucose content were lower in Co 1163 than Co 740. Different levels of K_2O do not significantly effect the glucose content of *gur*. This is perhaps due to the fact that potassium reduces the ill effect of nitrogen per cent in the cane juice. It maintains the balance of N : K and lowers the N : P_2O_5 ratio by increasing P_2O_5 and K_2O in cane juice (Misra *et al.*, 1964a; Singh *et al.*, 1983; Patil and Shingte, 1982a). Patil and Shingte (1982a) conclusively proved that different levels of potassic fertilizer do not influence the moisture content of *gur* (Table 6.6) and thereby confirmed similar observations reported by Misra *et al.* (1964).

Application of potassium at or above 112 kg K_2O ha^{-1} alone or in combination with N and P adversely affected cane yield and quality of *gur*, pol per cent juice, purity coefficient and invert sugars. Besides it also increased the ash content (Anon., (Shahjahanpur, India), 1983). According to Patil and Shingte (1982b) when press mud and 100 kg K_2O ha^{-1} were applied to cane, improvement in juice content was clearly discernible.

In an experiment conducted to find out response of sugarcane variety Co 740 to the application of potassium and phosphorus fertilization for improving the quality of *gur* under conditions prevailing in Kolhapur district of Maharashtra, during 1982-83 and 1983-84, Patil and Shingte (1992) also confirmed that application of 110 and 165 kg K_2O ha^{-1} significantly (i) increased sucrose content, (ii) reduced reducing sugars and (iii) improved colour but did not influence brix and moisture of *gur* over 0 and 55 kg K_2O ha^{-1} during both the seasons. At Kolhapur (Maharashtra), application of potash up to 200 kg K/ha and harvesting a crop of mid-late ripening variety (Co 7527) at the age of 14 months improved the quality of *gur* (Wandre and Hasabnis, 1991).

Effect of Phosphatic Fertilizers

Application of phosphorus not only improves the quality but also mitigates the adverse effect of high nitrogen dose. Foliar application of P did not influence *gur* quality, while its soil application showed improvement in juice claribility and the quality of *gur*. However, application of 20 kg P_2O_5 ha^{-1}, as single super phosphate (basal application), showed no significant improvement in *gur* quality (Anon., 1954). Under Shahjahanpur conditions, neither the response of phosphorus nor the deterioration in quality of *gur* was observed (Gupta and Singh, 1974). However, under Maharashtra conditions, a dose of 100 kg P_2O_5 ha^{-1} to sugarcane crop gave the best quality *gur* (Dagade and Ranadive, 1980). Application of 75 kg P_2O_5 and 112 kg K_2O ha^{-1} generally improved the yield and quality of *gur*. However, a dosage of 115 kg ha^{-1} each of P_2O_5 and K_2O is optimum for the preparation of quality *gur*. Patil and Shingte (1992) observed that either sucrose content or reducing sugars of *gur* was not influenced by the application of 0, 55, 110 and 165 kg P_2O_5 ha^{-1} at 10 and 11 months age and at harvest. It also did not influence colour intensity, brix and moisture content in *gur*.

At Melalathur (Tamil Nadu), application of 50-75 kg P_2O_5 ha^{-1} in the form of superphosphate as basal dose and top dressing of 112-150 kg K_2O as MOP on 4th and 90th day after planting improved yield of cane, juice quality and brought a reduction in reducing sugars, so also the yield and quality of *gur* (Chockalingam, 1985). Studies conducted at Shahjahanpur indicated that sugarcane grown under low (75 kg N ha^{-1} + 2 pre-monsoon irrigations) and medium (10 kg N ha^{-1} + 2-4 pre-monsoon irrigations) fertility status produced quality *gur* in comparison to high fertility (225 kg N ha^{-1} + 6 pre-monsoon irrigations) status (Lal *et al.*, 1991).

Effect of Micronutrients on Gur Yield

In micronutrient deficient soils at Melalathur (Tamil Nadu) foliar spraying of 1.5 per cent ferrous sulphate + 1 per cent zinc sulphate + 1 per cent urea at monthly intervals (from 2nd month to 10th month after planting) resulted in an increase in sugarcane yield (13%) and *gur* recovery (12.4%) over control (7.5%) (Shakunthala, 1991).

Effect of Press Mud on Quality of Gur

Press mud, a by product of sugar industry having sulphitation process, is generally added in alkaline soils to boost sugarcane productivity. Besides, it appreciably increases the water holding capacity of the soil, keeps the population of nematodes below the economic injury level and stimulates the growth of antagonistic microorganisms in soil.

Investigations by different workers clearly indicate that this factory waste is better than farm yard manure (FYM) or compost. Combination of press mud with inorganic N fertilizers not only boost cane productivity but also give superior quality of *gur*. Patil and Shingte (1982a) studied the role of press mud on production of *gur* its quality (Table 6.7).

Table 6.7: Effect of Different Levels of Press Mud on Quality of *Gur* from Sugarcane Variety Co 740

Level of Press* mud (t ha^{-1})	Sucrose (%)	Glucose (%)	Colour intensity (green filter)
7.4	79.85	5.04	103
11.1	79.47	5.75	106
14.8	78.12	5.91	119
18.5	77.47	6.15	123
SE ±	0.26	0.32	2.32
CD at 5%	0.78	0.95	6.70

* had basal dose of 112 kg P_2O_5 and 112 kg K_2O ha^{-1}.
Source: Patil and Shingte, 1982a.

They reported that maximum quantity of sucrose was recorded when 7.4 t of press mud was applied per ha. Further increase in quantity of press mud from 11.1 to 18.5 t per ha gradually decreased the quantity of sucrose content. The amount of glucose increased with an increase in press mud. The higher concentration of nitrogen with corresponding lower amounts of P_2O_5 perhaps brought down the purity of the juice as adjudged by an increase in the amount of glucose.

The colour of the *gur* also changed as the quantity of the press mud applied increased from 7.4 to 18.5 t ha^{-1}. At the highest concentration, the colour of the *gur* was reddish black. Kale (1957) had also reported dark coloured *gur* because of the abundance of salts in the juice.

Manuring and Irrigation of Ratoon in Relation to *Gur* Quality

Growing of sugarcane ratoon under rainfed conditions with full dose of nitrogen (67.6 kg ha^{-1}), applied at the break of monsoon resulted not only in low recovery of *gur* per cent cane and yield of *gur* but also deteriorated its quality. Most suitable interval between pre-monsoon irrigations was about 15 days. Increase in N dose influenced recovery of *gur*, determined as per cent of juice or cane, when irrigated at 15 days interval, while reverse results were observed when irrigation was given at 21 days interval. Pre-monsoon irrigation (1-3) resulted in a marked increase in *gur* yield per hectare (Gupta and Singh, 1974). However, application of N in single or in two splits along with one pre-monsoon irrigation did not show any marked effect either on yield or quality of *gur*. Higher yield of ratoon cane (variety Co S 510) obtained by application of 150 kg N ha^{-1} with 4 pre-monsoon irrigations a higher yield of *gur* as compared to 150 kg N ha^{-1} with 3 irrigations and 200 kg N ha^{-1} combined with 5 pre-monsoon irrigations (Anon., (Shahjahanpur, India), 1983).

Effect of Weeds on Quality of Juice

In some places, it has been observed that the *gur* made from canes obtained from a field full of weeds was relatively inferior than the one obtained from a weed free field. Weeds are known to remove up to 44 kg P_2O_5 ha^{-1} (Mani, 1975) which disturbs the N/P ratio. This may affect the juice quality and ultimately the quality of *gur* (Tembhekar, 1954).

Table 6.8: Effect of Pesticides (Fungicides and Insecticides) on Yield of *Gur*

Treatment	Yield of gur (t ha^{-1})
Aretan	4.80
Chlordane	4.27
Aldrin	4.52
Gamma HCH	5.21
Heptachlor	5.33
Aretan + Chlordane	5.15
Aretan + Aldrin	5.79
Aretan + Gamma HCH	5.57
Aretan + Heptachlor	7.37
Control	2.91

Source: Misra and Singh, 1964.

Effect of Applying Chemicals at Planting on Gur Yield

Misra and Singh (1964 b) have studied the effect of fungicides (organo-mercurial compounds) which improve and enhance germination and some insecticides which are effective in controlling shoot borer/termites on crop stand and final yield and recovery of *gur* (Table 6.8).

Yield of *gur* in the treated crop, in general, and heptachlor + aretan, in particular, was much higher than in the control (untreated check). The final yield of *gur* per ha worked out to be 70 per cent higher due to heptachlor + aretan treatment. By and large, all the combined treatments proved superior to all the other treatments in which insecticides were used singly.

Gur Recovery in Tannery Effluent Effected Soils

Studies on tannery effluent affected soil showed that the *gur* prepared from sugarcane grown on such soils had less sucrose (86.6 per cent) as compared to normal soil (90.3%). The ash content and moisture per cent was found to be more in *gur* from effluent affected soil. The effluent affected soil produced dark brown coloured *gur,* while it was pale yellow from cane growing in normal soils (Devraj and Shakunthala, 1991).

REFERENCES

Agrawal, M.P. and A.K. Ghosh (1983). Effect of cane maturity, nitrogen fertilizer, clarificants and additives on keeping quality of *gur. Indian Sugar* 33(2): 1-8.

Anonymous (1877-78). *Gazetteer of Oudh (1877-78),* reprinted in 1985, B.R. Publication Corporation, Delhi, vol. III, pp. 80, 357, 520.

Anonymous, (1959). Sugarcane Research in India: 1929-1954 : Gur and Khandsari. *Indian J. Sugarcane Res. and Devpt.* III (pt. 2): 114-122.

Anonymous (Anakapalle, India) (1964). *Annl. Rept.,* Regional Sugarcane Research Station, Anakapalle, Andhra Pradesh.

Anonymous (Anakapalle, India) (1989). 75 Years of Agriculture Research at Anakapalle, Regional Research Station, Anakapalle, APAU, Hyderabad, India, pp. 400.

Anonymous (1954). *Annl. Rept.,* Sugarcane Research Station, Pusa, Bihar, pp. 30-31.

Anonymous, (Shahjahanpur, India) (1983). 70 Years of Sugarcane Research (1912-1981). U.P. Council of Sugarcane Research, Uttar Pradesh, pp. 59.

Asokan, S., S. Lalitha, S. Thangavelu and K. Chirajnivi Rao (1978). Urea vs. ammonium sulphate on juice composition and jaggery quality. *Proc. D.S.T.A.* 28(1): 54-58.

Barnes, J.H. (1918). Sugar and the Sugarcane in the Gurdaspur district. Agricultural Research, Pusa, Bihar, Bull. No. 69, Supdt., Govt. Printing India, Calcutta, pp. 100.

Chockalingam, S. (1985). Gur manufacture in Tamil Nadu—A Review, Proc. *National Seminar-cum-Group Discussion on Jaggery Manufacture and Storage* Dec., 18-19, Indian Institute of Sugarcane Research, Lucknow, India, pp. 94-98.

Deerr, N. (1931). (*c.f.* Narasimham, 1957).

Devraj, G. and V.A. Shakunthala (1991). Studies on the quality of *gur* prepared from cane grown in tannery effluent affected soils, paper presented at the *National Seminar on Increasing Productivity, Quality, Storability of Jaggery (Gur),* organised by the Indian Institute of Sugarcane Research, Lucknow and Indian Society of Agricultural Engineers, New Delhi held at the Indian Institute of Sugarcane Research, Lucknow, Nov., 30-December, 1, 1991.

Dagade, V.G. and S.J. Ranadive (1980). Effects of N and P fertilization for keeping quality of jaggery of different sugarcane varieties. *Maharashtra Sugar,* 5(12): 51.

Dagade, V.G. and S.J. Randive (1988). Effect of soil characters and irrigation on quality of *gur*. *Bharatiya Sugar,* 13(12): 39.

Gupta, A.P., R.S. Patil and V.K. Gupta (1986). Effect of press mud and zinc on the yield and uptake of zinc and nitrogen by corn. *J. Indian Soc. Soil. Sci.,* 34(4): 810-814.

Gupta, D.N. and R.Singh (1974). Response of phosphate fertilization as influenced by bulk organic matter and irrigation levels. *Co-op Sugar,* 5(6): 274-277.

Hadi, S.M. (1929). The Indian Sugar Industry, Govt. of Bhopal, C.I., printed by F.W. Petch Thacker and Co. Ltd., Bombay, published by Khan Bahadur S.M. Hadi, Director Agriculture, Bhopal State, pp. 289 + vi.

Hoda, M.M. (1985). Sessional Reports in *Proc. National Seminar-cum-Group Discussion on Jaggery Manufacture and Storage,* held at Indian Institute of Sugarcane Research, Lucknow, India, Dec., 18-19, pp. 154-161.

Javelekar, D.V. (1979). Studies on the effect of different juice clarificants on the physio-chemical properties and keeping quality during storage of jaggery, M.Sc., (Agric.) thesis, Mahatma Phule Agricultural University, Rahuri, Maharashtra, India.

Kale, R.A. (1957). Sugarcane Cultivation in Bombay State, Central Sugarcane Research Station, Padegaon, pp. 131-139.

Khanna, K.L. and A.S. Chacravarti (1949). Studies in clarification of sugarcane juice in *gur* manufacture. *Indian J. Agric. Sci.,* 19(1): 137-161.

Krishnamurthy Rao, K. (1919). The effect of salinity on the growth and composition of sugarcane. *Agric J. India,* 14: 476-493.

Lal, U., S. Singh and S.C. Srivastava (1991). Effect of agronomic practices on gur quality, paper presented at the *National Seminar on Increasing Productivity, Quality, Storability of Jaggery (Gur),* organised by the Indian Institute of Sugarcane Research, Lucknow and Indian Society of Agricultural Engineers, New Delhi held at the Indian Institute of Sugarcane Research, Lucknow, Now., 30-December, 1, 1991.

Lender, P.E. and R. Narain (1936). Mineral matter in the juice of sugarcane, its effect on recovery of white sugar. *Indian J. agric. Sci.* 6(4): (*c.f.* Narasimham, 1957).

Lugo-Lapez, M.P. (1954). The effect of soil condition on the sucrose content of sugarcane. *J. Agic. Univ. Puerto Rico,* 38: 32-46.

Mani, V.S. (1975) (*c.f.* Tandon, 1987).

Misra, G.N., R.C. Kapoor and S.C. Johari (1964 a). Effect of N, phosphate and potash on yield and quality of *gur. Indian Sugar,* 9(1): 32-39.

Misra, G.N. and J.D. Singh (1964 b). Improving crop stand and final yields of sugarcane and *gur* through the application of chemicals (fungicides and insecticides). *Proc. Bien. Conf. S.R.D.W.I.,* 5: 119-131.

Misra, G.N., B.S. Mathur and J.D. Singh (1965). Response of sugarcane to different nitrogenous fertilizers, effect on yield of cane and gur and its quality. *Proc. Ann. Conv. Sugar Technol. Assoc. India*. 33: Ag. 39.

Mukherjee, S.R., U.S. Prasad, C.R. Prasad and B.P. Shai (1988). Studies on the *gur* characteristics in relation to soil properties of some typical *gur* making areas in Bihar. *Bharatiya Sugar*, 13(3): 27.

Narasimham, R.L. (1957). Regional variation in the composition of sugarcane juice in relation to claribility. *Proc. Annl. Conv. S.T.A.I.*, 25: 27-42.

Patil, J.P. and A.K. Shingte (1982a). Studies on the effect of different levels of press mud, K and nitrogen on the quality of jaggery. *Indian Sugar*, 32(3): 167.

Patil, J.P. and A.K. Shingte (1982b). Effect of potassium fertilization and sugarcane varieties on quality of jaggery. *Indian Sugar*, 32(6): 377.

Patil, J.P. and A.K. Shingte (1992). Effect of application of phosphorus and potassium to sugarcane crop and its effect on quality of jaggery. (Part I.) *Indian Sugar*, 41(2): 385-392.

Prasad, K. and D.P. Srivastava (1954). Effect of manurial and cultural treatments on claribility of sugarcane juice in open pan boiling process I. Nitrogen, phosphate manuring and irrigation. *Proc. Bien. Conf. S.R.D.W.I.*, 2(11): 584-591.

Ranadive, S.J. (1985). Review of the work done on *gur* at Sugarcane Research Station, Padegaon. *Proc. National Seminar cum Group Discussion on Jaggery Manufacture and Storage*, held at the Indian Institute of Sugarcane Research, Lucknow, Dec., 18-19, pp. 99-104.

Rege, R.D. and V.V. Tembhekar (1954). Studies on the influence of brix and purity of juice on the production of gur. *Proc. Bien. Conf. S.R.D.W.I.*, 2(11): 572-574.

Shakunthala, S.V. (1981). *Manual of Sugarcane Production Technology* Department of Agriculture, Tamil Nadu, pp. 142-143.

Shakunthala, V.A. (1991). Studies on the effect of micronutrients on cane yield and *gur* recovery in different soils, paper presented at the *National Seminar on Increasing Productivity, Quality, Storability of Jaggery (Gur)*, organised by the Indian Institute of Sugarcane Research, Lucknow and Indian Society of Agricultural Engineers, New Delhi held at the Indian Institute of Sugarcane Research, Lucknow, Nov., 30-December, 1, 1991.

Singh, M., R.S. Mathur and R.K. Sharma (1983). *Gur* quality attributes, effect of nitrogen, phosphate and potash. *Co-op. Sugar*, 14(7): 359-362.

Swadi (1919) *c.f.* Anonymous (1959).

Tandon, H.L.S. (1987). Phosphorus Research and Agriculture Production in India. Fertilizer Development and Consultation Organisation, New Delhi, pp. 160.

Taylor, C.S. (1913). *Notes on experiments with sugarcane at Sabour Pt. I: On the character and quantity of fibre in different varieties, Pt. II: On the effect on different manurial treatments on the time of ripening of cane.* Agricultural Research Institute, Pusa (Bihar), *Bulletin No.* 37, pp. 17, Supdt. Govt. Printing, India, Calcutta.

Tembhekar, V.V. (1954). Effect of mineral ingredients present in the sugarcane juice on the quality of gur. *Proc. Annl. Conv. D.S.T.A.*, 11(1): 161-163.

van Dillewijn, C. (1948). Maturity Testing, *Bull. No.* 3. *Sugarcane Expert, Turkist Sugar Factories, J.S.C. Ankara*, pp. 111.

Wandre, S.S., J.R. Patil and A.K. Shinde (1985). Effect of nitrogen levels on quality and storability of jaggery (Co 740). *Proc. National Seminar cum Group Discussion on Jaggery Manufacture and Storage,* held at the Indian Institute of Sugarcane Research, Lucknow, December, 18-19.

Wandre, S.S. and A.B. Hasabnis (1991). Effect of different potash levels and harvesting dates on quality of jaggery, paper presented at the *National Seminar on Increasing Productivity, Quality, Storability of Jaggery (Gur),* organised by the Indian Institute of Sugarcane Research, Lucknow and Indian Society of Agricultural Engineers, New Delhi held at the Indian Institute of Sugarcane Research, Lucknow, Nov., 30-December, 1, 1991.

7

CLARIFICANTS USED IN THE MANUFACTURE OF *GUR*

The sugarcane juice is an opaque liquid and its colour varies from grey to dark green. The raw juice contains soluble substances, such as, sucrose, glucose, fructose, amino acids, organic acids, etc. It also contains suspended materials like colloides, wax, colouring matter, albumin, gum and *bagacillo*. These suspended substances, if allowed to remain in juice, impart a disagreeable colour and taste to the *gur*. Therefore, it becomes imperative to remove them from juice as far as possible. The process of clarification also aims at preventing the formation of non-sugars during heating due to the acidity of juice. Removal of these can effectively be achieved by using vegetable clarificants, certain chemicals or both.

Clarificants used in the manufacture of *gur* may be classified into two broad categories:

Vegetable Clarificants

As early as in 1893, Watt reported the use of ash of certain plants for clarification of cane juice. In 19th century, 'rice flour' was used for refining raw sugar in India and China. In the manufacture of *gur* by open pan process a number of vegetable clarificants are used to facilitate removal of scum from raw juice (Bharadwaj *et al.*, 1976; Javelekar *et al.*, 1985; Vaidya *et al.* 1984).

At present, vegetable clarificants used for the clarification of cane juice are: *deola (Hibiscus ficulneus)*, *bhindi* (*Abelmoschus esculentus*), *sukhlai* (*Kydia calycina*), *semal* bark (*Bombax melabaricum*) and *falsa* (*Grewia asiatica*), and extract of groundnut (*Arachis hypogaea*) and castor (*Ricinus communis*) seeds. A vegetable clarificant should be preferably be edible and should not affect the taste of *gur*. Vegetable clarificants used for *gur* manufacture along with the quantity required for clarifying 100 litre juice are given in Table 7.1.

Table 7.1: Vegetable Clarificants used for Juice Clarification in the Manufacture of *Gur*

Clarificant	Quantity used g/100 kg juice	Reference
Extract of *Deola* (*Hibiscus ficulneus*)	37.5	Agarwal *et al.*, 1987
	56.2	Anon. (Shahjahanpur, India), 1983
	37.5+ 10 g hydros	Anon. (Shahjahanpur, India), 1983
	2.3 (0.5 kg *deola* per 100 L water	Chockalingam, 1985
Seed extract of groundnut (*Arachis hypogaea*)	40 g	
Seed extract of caster (*Ricinus communis)*		
Bhindi mucilage of (*Abelmoschus esculentus*)	50 g	Agarwal *et al.*, 1987
	*1.14 – 2.86 L + super phosphate	Ranadive, 1985
Semal (Bombax melabaricum)	62.5 g	Agarwal *et al.*, 1987
Falsa (Grewia asiatica)	56.2 g	Agarwal *et al.*, 1987
Sukhlai (Kydia calycina)	50.0 g	Agarwal *et al.*, 1987
	62.0 g + usual liming	Anon., (Anakapalle, India), 1989
	61.0 g	
Soybean flour + monocalcium chloride (can be made into tablets)	42.7 g+ 0.93 g	Khanna and Chacravarti, 1955
Hibiscus vitifolius	—	Shakunthala, 1985
**Kasthuri (H. abelmoschus* or *Bamia moschus)* tender leaves and roots)		Shakunthala and Devraj, 1991

*Superior clarificant than *Deola* or *Bhindi* etc.

Preparation of Commonly used Clarificants of Vegetable Origin

Deola (Hibiscus ficulneus)

This plant belongs to the family Malvaceae and is the most commonly used vegetable clarificant. It is prepared by crushing stem and root of the plant in a small quantity of water. After some time the crushed material is rubbed rigorously in water till a thick mucilagenous liquid is obtained in sufficient quantity. The whole process is repeated several times in order to obtain more and more of thick mucilage. When no mucilage comes out of the crushed material it is discarded. About one and a half kg of the plant material is required for clarifying 3.5 tonnes of juice.

Bhindi (Hibisucs esculentus)

It belongs to the family Malvaceae and is commonly known as 'Lady's finger'. It is used in the similar fashion as described for *deola* but is of slightly inferior quality as compared to *deola* because of less production of mucilage. About 1.75 kg of *bhindi* plant is used for clarifying 3.5 tonnes of raw juice.

Sukhlai (Kydia calycina)

It also belongs to the family Malvaceae. The bark of the plant, green or dry, is soaked in sufficient quantity of water overnight. Soaking is a prerequisite for the extraction of mucilage. The bark is rubbed vigorously and squeezed to extract the mucilagenous substance. This process is repeated many times till no extraction of mucilage is there. About 1.5 kg of *sukhlai* bark is needed for the clarification of 3.5 tonnes of juice.

Semal (Bombax melabaricum)

This plant also belongs to the family Malvaceae. Only its green bark is crushed in mortar and is kept dipped in water for 2-3 hr and rest of the procedure is identical to that described for *sukhlai*. About 2 kg of green bark is needed for clarifying 3.5 tonnes of juice.

Falsa (Grewia asiatica)

It belongs to the family Tiliaceae. Here also green bark of the plant is used in the identical fashion as outlined for *semal*. About 1.75 kg of the bark is needed for the clarification of 3.5 tonnes of raw juice.

Groundnut (Arachis hypogaea)

This plant belongs to the family Leguminosae and because of its easy availability is used throughout the length and breadth of the country. Groundnuts are first de-selled to remove kernels. These are crushed in a mortar with water. The resultant milky liquid is passed through a muslin cloth or a wire mesh and collected in a pot. This process is repeated several times till the kernels yield milky white liquid. This milky white liquid is added to the cane juice while it is being heated. The scum that comes to the surface is frequently removed with the help of a laddle. Extract from 2.5 kg kernels is sufficient to clarify 3.5 t of raw juice. It is generally used for clarifying juice of immature or deteriorated cane as also from canes grown under waterlogged conditions. Bharadwaj *et al.* (1976) found that excess of this clarificant adversely affected the taste and quality of the *gur.*

Castor (Ricinus communis)

This plant belongs to the family Euphorbiaceae. The seeds of castor are crushed in a mortar with water as mentioned for groundnut and the method of use is also the same. About 2.5 kg of castor seed is used for the clarification of 3.5 tonnes of sugarcane juice.

Castor is one of the good vegetable clarificants used in India and its use is generally recommended when manufacturers want desirable yellowish colour and fair degree of hardness of *gur.*

Kasthuri (Hibiscus abelmoschus or Bamia moschus) or Muskmallow

Tender leaves and roots of *kasthuri* have also been used as a clarificant for sugarcane juice. This plant contains gums, albumin, essential oils etc. and its seed contains aromatic essential odour resembling musk. It has been found to be a better clarificant than *deola* and *bhindi* (Shakunthala and Devaraj, 1991).

As regards the comparative effectiveness of different clarificants mentioned above they can be mentioned in a descending order: *kasthuri, deola, bhindi,* groundnut, *semal, falsa, sukhlai* and castor seeds. No work on clarificants has been done in recent years.

In North India, vegetable clarificants (Table 7.1) have been found to be better than chemical ones but in South India, soft *gur* is produced using only vegetable clarificants. The main reason for this is that cane juice in South India has high acidity (pH 5.2–5.3), therefore, use of lime for clarification is a must with all vegetable clarificants.

Chemical Clarificants

In addition to the clarificants of vegetable origin mentioned above, chemical ones are also employed to supplement the efficacy of the former; this is usually done with a view to imparting better texture and brilliance to the *gur*. It has been observed that the effect of these chemicals is not long lasting because after a couple of days the attractive colour fades away leaving the product worse than the original.

In recent years, the use of chemical clarificants has been generally avoided as far as possible. In India the *gur* is priced for its good flavour and taste and often the flavour is lost due to the action of these chemicals, resulting in reduced market value. Furthermore, these chemicals introduced inorganic salts in *gur*. These being hygroscopic in nature reduce the keeping quality of *gur*. It is because of these reasons, the use of chemicals is not advised. At the same time it is advisable to use chemicals like lime water or soda when the juice is of inferior quality and the *gur* does not set well.

In India, the following chemical clarificants are generally used:

Lime

It is the most widely used chemical clarificant because it is cheap, easily available and clarifies the juice effectively. Lime also serves as a preservative in preventing inversion of juice till boiling begins. Care is necessary to ensure correct neutralisation of the acidity of the juice. For obtaining good colour and hardness liming to pH 6.3–6.6 is essential. Inadequate liming makes the *gur* soft but imparts a good colour. Conversely, excess liming makes the *gur* hard and imparts a dark colour to it. Experience has shown that liming to pH 6.4 is the best for juice obtained from immature cane and pH of 6.8-7.2 for the juice obtained from deteriorated canes having purity not above 75 per cent. It may be mentioned that excessive liming at higher temperatures should be avoided as it leads to dark coloured *gur*.

Table 7.2: Use of Clarificants in Making *Gur* from Deteriorated Canes or Crops Grown under Abnormal Conditions

Condition	Clarificant used	Reference
Immature, water-logged and deteriorated canes	Milky extract of castor seeds	Vaidya *et al.*, 1984
Immature, mature and deteriorated canes	Blanching in hot water at 80-90□C for 20 minutes	Prasada Rao *et al.*, 1984
Immature, water-logged/ termite infested canes	Castor seed and groundnut extract	Javelakar *et al.*, 1985
Stand over canes (beyond May)	Addition of lime	Chockalingam, (1985)
Lodged canes	*Deola* extract 37.5 g followed by 5% *Sajji* (0.25-0.38 l/q of juice)	Lal *et al.*, (1985)
Poorly managed canes	*Deola* extract + 10 g hydros	Aagarwal *et al.*, (1987)
Acidic juice (pH 5.5)	25 g lime (80-90 % purity) to bring pH slightly below 6.4	Chockaligam, (1985)
Early ripening varieties	37.5 g *deola* estract + 10 g hydros	Agarwal *et al.*, (1987)
Immature canes (10-month age)	Lime alone or Lime sucrate (pH 6.4)	Anon., (Anakapalle, India) (1989)
Mature cane (12-month age)	Lime alone or Lime sucrate (pH 5.8)	-do-
Deteriorated canes (14-month age)	Lime alone or Lime sucrate (pH 6.6) + some *gur* powder	-do-

Lime sucrate is prepared as follows: One part of lime is slaked with and more water is added till the proportion of water to lime is 5 : 1. The solution is mixed well and strained through a coarse cloth to get lime water. One part of lime is stirred in 5 parts of cane juice and stirred and kept over night. Next day the clear supernatant liquid is decanted and used.

Sodium Bicarbonate

It is generally used with vegetable clarificants particularly with *deola (Hibiscus ficulneus)* to temporarily improve the *gur* colour. Use of sodium carbonate adversely affects the *gur* quality. It is used only when *gur* is prepared from inferior quality cane.

Sajji (Saline-alkaline Soil)

Composed of 50 per cent sodium carbonate + 6.4 per cent sodium sulphate and 4.5 per cent sodium chloride and is generally employed for making *gur* from unripe canes but the taste of the finished product is spoiled.

Superphosphate

Use of this chemical imparts a golden colour to *gur* but its crystalline structure is lost. Its keeping quality is also poor.

Alum

Alum is generally used for making semi-solid *gur* of brilliant light colour after the usual clarification with some vegetable clarificant like *deola.* As the chemical has an acidic reaction, it inverts the sucrose in juice at the time of boiling and does not allow the *gur* to solidity in crystalline form.

Sodium Hydrosulphite

Commonly known as 'hydros' and is the most commonly used clarificant in recent years. It bleaches the colour of *gur* and as a result there is appreciable improvement in *gur* colour (light yellowish). The only disadvantage with this is that colour improvement is temporary and the product deteriorates after some time. Another undesirable effect of hydros is that it is injurious to health because of high SO_2 content in the *gur.* According to Prevention of Food Adulteration Rules, 1955 *gur* should not contain 70 ppm of SO_2. Use of even 0.25 g of hydros per litre increases the SO_2 content of *gur* above 75 ppm, although at this hydros level no improvement in colour is obtained. So in actual practice, manufacturers are using much higher concentration of hydros.

It is true for all the chemical clarificants that these are used in small amounts in boiling juice after the same has been clarified with some vegetable clarificant.

In immature canes, phenolic contents are relatively more. Addition of 50 ppm of P_2O_5 reduces the total phenolics by 40.23 per cent; however, further addition of P_2O_5 (125 ppm) reduces it only by 45.0 per cent. Addition of lime also eliminates phenolics; this is dependent on the pH of the medium. At 6.8 pH, 35.8 per cent of the total phenols were removed, whereas at 7.4-8.2 pH reduction of phenolic content was relatively less (Sharma *et al.*, 1980).

Use of Artificial Colouring Agents to Impart Colour

For improving colour of *gur,* Dagade and Ranadive (1983) compared the addition of two dyes, namely, tetrazine and erythrosin with hydros after removal of scum by mucilage of *bhindi.* They reported that Tetrazine was *at par* with hydros as regards the quality of *gur* obtained from Co 775 and Co 62175, at Padegaon (Maharashtra).

Use of Flocculents

The flocculation is a process where finely divided or colloidally dispersed particles in solution gather together to form big particles of such a size that settle early or it is an agglomerate of tiny particles to form 'flocs' which easily settle and enhance the clarification of the juice.

Substances that cause flocculation are known as flocculants or flocculating agents. The choice of flocculant depends upon various factors which one comes across during the process such as viscosity, colloidal contents, temperature, pH of the medium, chemical nature of the medium, incompatibility, etc.

The optimum concentration of the flocculant is that which gives a good coagulation of colloidal non-sugars for obtaining a clear juice of the highest purity.

Flocculants, synthetic high polymers, are polyanions, polyacrylates, polyacrylamides, partially hydrolised polyacrylonitriles or co-polymers of vinylacetate and maleic hydride. They exist in the form of thread like long chains and the molecular weight in the region of 3000 is necessary. The molecule has to be long enough as it has to bridge the gap between the two adjacent molecules of suspended matter. Flocculant should be high polymers which behave as colloidal electrolytes, that is with the large organic chains with ionizable group attached at the end.

Chaterjee *et al.* (1975) advocated the use of flocculants, 7 ppm of Primofloc A-10, Flocal L.T. 26/53 (NCL) and 10 ppm of flocculants 4, 5 and 6 (NCL) and Indole to juice adjusted to pH 7.2 for better result in settling the non-sugasr impurities as mud. Mukherjee *et al.* (1978) have compared the efficacy of the electrolyte TF-S_3, a product of True Food Corporation (India), as a flocculating agent against Separan AP 30 (Dow Chemical, U.S.A.) as standard. Study revealed that sugarcane juice when treated with five gram lime + 15 g single super phosphate and 10 ml of 0.1 per cent TF-S_3 per litre of juice was as good as Separan AP 30. Extract of *Piper betel;* however, proved to be an inferior clarifying agent.

REFERENCES

Agrawal, M.L., U. Lal, R.K. Sharma and S.K. Varma (1987). Effect of variety, standards of cultivation and juice clarificants on *gur* quality parameters. *Bharatiya Sugar* 12 (11) : 33.

Anonymous (Shahjahanpur, India) (1983). *70 Years of Sugarcane Research (1912-1981).* U.P. Council of Sugarcane Research, Uttar Pradesh, pp. 59.

Anonymous (Anakapalle, India) (1989). *75 Years of Agricultural Research (1913-1988) at Anakapalle,* Regional Research Station, Anakapalle, A.P.A.U., Hyderabad, Andhra Pradesh, pp. 400.

Bharadwaj, K.M., M. Singh, and R.K. Sharma (1976). Practical importance of clarification in *gur* industry. *Sugar News (India)* 7(11): 19.

Chatterjee, A.C., S.R. Kalaswad, M.R. Math and V.P. Taware (1975). Evaluation of flocculating agents in cane juice clarification. *Indian Sugar* 25(9): 723-727.

Chockalingam, S. (1985). *Gur* manufacture in Tamil Nadu—A Review. *National Seminar-cum-group discussion on jaggery manufacture and storage,* held at IISR, Lucknow (Dec. 18-19).

Dagade, V.G. and S.J. Ranadive (1983). Effect of different edible colour agents on keeping quality of jaggery of different varieties. *Maharashtra Sugar* 6(1) : 123.

Javalekar, D.V., B.N. Shide and S.J. Ranadive (1985). Use of different clarificant in jaggery making. *Indian Sugar* 36(6) : 275.

Khanna, K.L. and A.S. Chacravarti (1955). Scientific Monograph—Researches on Technical Aspects Relating to Improvement to *Gur* Industry in Bihar, Indian Central Sugarcane Committee, New Delhi.

Lal, U., S.C. Srivastava, R.K. Sharma and S.K. Varma (1985). Studies on the improvement of keeping quality of *gur* prepared from lodged cane. *Proc. National Seminar cum Group Discussion on Jaggery Manufacture and Storage* held at IISR, Lucknow, December 18-19.

Mukherjee, S.R., R.D. Sahi and B.P. Sahi (1978). Utilization of TF-S_3 as a clarifying agent in *gur* industry. *Proc. Ann. Conv. Sugar Technol. Assoc. India,* 22: M 13-18.

Prasada Rao, K.K., K. Ramalingaswamy, P.V.L.N. Raju and G. Vishwanathan (1984). Studies on jaggery from hot water treated immature, mature and deteriorated cane. *Maharashtra Sugar* 9(4): 39.

Shakunthala, V.A. and G. Devraj (1991). *Hibiscus abelmoschus* Linn.—A vegetable clarificant in *gur* making, presented at the *National Seminar on Increasing Productivity, Quality and Storability of Jaggery (Gur),* Organised by the Indian Institute of Sugarcane Research and Indian Society of Agricultural Engineers, New Delhi, held at the Indian Institute of Sugarcane Research, Lucknow, India, Nov., 30 to Dec., 1, p. 24.

Shakunthala, V.A. (1985). *Hibiscus vitifolius*—A new clarificant in the preparation of jaggery. *Proc. Seminar Cum Group Discussion on Jaggery Manufacture and Storage,* held at Indian Institute of Sugarcane Research, Lucknow, Dec., 18-19, pp. 90-93.

Sharma, S.C., P.C. Johari and G.S.C. Rao (1980). Fate of phenolic content during clarification of juice. *Proc. Intern. Conf. Sugar Cane Technology,* 17(III): 2137-2143.

Vaidya, V.R., A.D. Kadlage and D.G. Hapase (1984). The efficacy of different vegetable clarificants of quality jaggery preparation. *Maharashtra Sugar* 9(4): 69.

Watt, G. (1893). *The Dictionary of the Economic Products of India.*

8

GUR AND OTHER INDIGENOUS SUGARS: MANUFACTURE AND STRUCTURE

Process for the manufacture of *gur* consists of extracting the cell sap by passing the sugarcane stalks through a 3 or 5 roller crusher (vertical or horizontal) drawn by animal power or electric motor or by an oil engine. The juice should be boiled immediately after extraction in open pan by using sun dried bagasse and trash, as fuel. If the juice is not handled quickly, microorganisms develop and consume the sugar and after some time fermentation of juice starts. The boiling must be quick and vigorous, so that time taken by free acids, present in the juice, to invert the sucrose may be minimised. Juice obtained from the crusher is first strained through a piece of cloth to remove dirt, small bagacillo and any other extraneous material(s) suspended in it. During boiling, vegetable mucilaginous substances are added to remove the suspended impurities (as scum) by means of *pauna.*

The exact point at which the boiling is regarded as complete and the boiling mass is fit to be taken out of the pan for cooling is called the 'striking point'. When the boiling mass is near the required consistency i.e. 'striking point', the firing of the furnace is slowed down. The boiling mass is stirred frequently by means of a wooden stirrer in order to avoid charring due to excessive local heating.

For manufacturing of *gur,* skill or the artisan (juice boiler) plays an important role. An experienced worker can judge the exact consistency when the boiling mass has to be taken out of the pan. Over boiling results in the caramelisation of sugars affecting both flavour and colour of the *gur*. Conversely, underboiling prevents the *gur* from setting and solidifying. The striking point for *gur* usually lies between 110° and 120°C, depending on the quality of juice and type of *gur* to be manufactured. It is judged by the eye and never by a thermometer. The test described below gives an idea for determining the striking point during the preparation of *gur*:

(a) When boiling has progressed sufficiently, a small mass is taken out form the boiling pan using a wooden stirrer and immersed in cold water. The mass is then rolled into a ball. When the ball is hard enough to produce a metallic sound when thrown against the side of the pan, it is presumed that the boiling is complete. The boiling mass should immediately be taken out for cooling.

(b) At striking point, the boiling mass when stirred with a wooden stirrer will leave the bottom of the pan clear instead of sticking to it.

(c) If the wooden stirrer is taken out of the boiling mass and held out in air, the mass adhering to the stirrer will trickle down breaking into drops. When right consistency of the boiling mass is reached, the trickling mass will form long dry silky threads flowing in the air instead of breaking into drops.

(d) The striking point of the boiling mass is also judged by holding the mass between the thumb and index finger. On separating the thumb from index finger, the mass forms a long dry thread.

After the boiling mass has reached the striking point, it is transferred from the boiling pan to a shallow and circular iron pan or wooden tray called *chak* (cooling pan) (Fig. 8.1). The mass is allowed to cool up to the point when the upper layer has just solidified. At this stage, the mass is agitated vigorously by means of a *khurpi* (wooden scrapper) and is left in the *chak* to cool further. The semi-solid mass is then finally scrapped out of the cooling pan and is given the desired shape by hand or by transferring the mass to suitable moulds. This product is known as *gur* and can be made into different shapes ranging from small lumps of 10 g each to the bucket shape of 10 kg each.

Manufacture of *Gur* using Sugar Factory Machinery

Besides being made as an 'artisan sugar', at some places its manufacture is also taken up by some sugar factories (Anon., 1989).

The process of manufacturing *gur* by the Ugar Sugar Works (Maharashtra) using the sugar factory machinery (Anon., 1989) is shown in Fig. 8.2.

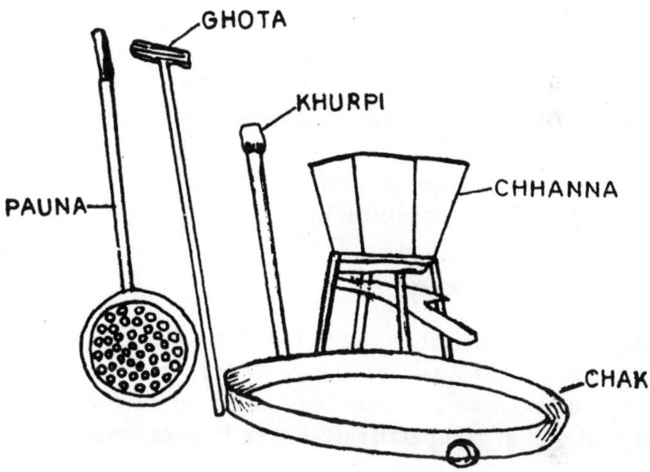

Fig. 8.1: Implements used in the Manufacture of *Gur*

The mixed juice is taken to a number of eliminators and coarse scum is removed by slowly bringing the juice to boiling point. After removing the scum, the juice is treated with the required amount of lime, phosphoric acid and Sodepur flocculent and is allowed to settle in tanks. The clear juice is siphoned off and evaporated to 'syrup' stage. The mud is filtered, and clear filtrate is pumped into a tank. Afterwards, it is sent to evaporators for further concentration. The syrup is then transferred to vacuum pans and concentrated to *gur* stage. About 250 to 300 blocks of 30 kg each are manufactured in one such operation. The colour, hardness and even sweetness are comparable to *gur* produced by the local *gur* manufacturing units.

The flow chart for manufacture of *gur* in the sugar factory is given in Fig. 8.2.

Estimates of Losses in Gur Making

Reviewing the performance of sugar industry in the earlier part of this century, Barber (1919) observed "the principles of *gur* making are undoubtedly in many places woefully wasteful". At Gurdaspur (now in Punjab State), Barnes (1918) gave estimates of losses in *gur* making from 34 to 57 per cent. This could be apportioned into 16 to 23 per cent due to bad milling, 13 to 14 per cent due to inversion in boiling juice and 1 to 22 per cent are invariably unaccounted losses perhaps due to destruction of sugar in boiling, lower milling efficiency, losses in handling raw juice, etc.

Use of Anti-foaming Agents

Pandit and Singh (1983) suggested the use of anti-foaming agents like POEM Z 001 to reduce the caramelisation during boiling, to increase *gur* recovery and fuel economy due to setting of *gur* at comparatively lower brix of juice.

Other Indigenous Sugars made from Sugarcane Juice

In Gurdaspur, in the early part of this century, the following sugar products were made:

- (a) *Gur* - a raw sugar made by simple evaporation of the juice
- (b) *Shakkar* - a light dry *gur*
- (c) *Rab* or *mal rab* - a liquid sugar
- (d) *Khand* - a refined sugar from which molasses has been drained off
- (e) *Bura* - made from *khand* boiled in water and clarified with milk

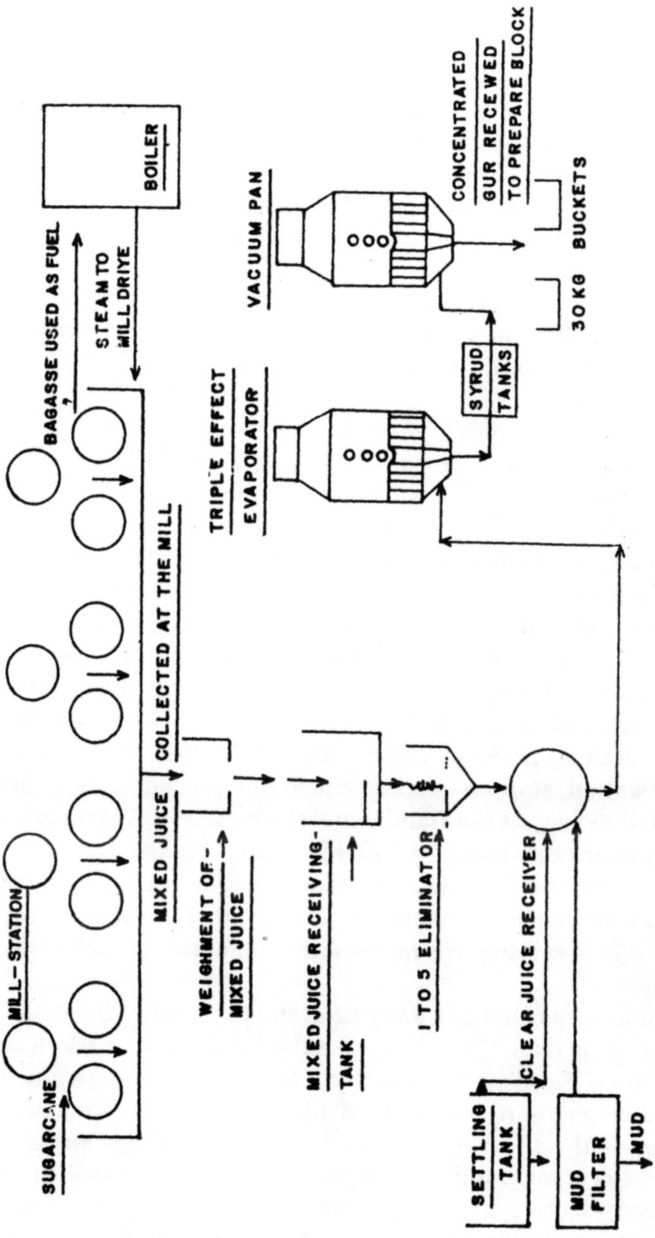

Fig. 8.2: Flow Chart for the Manufacture of *Gur* in the Sugar Factory

(f) *Misri* - prepared from *khand* boiled with water and evaporated

(g) *Kuza misri* - prepared as *misri* only with the best type of *khand*. It is prepared using earthen pots (*kuza*) placed in water. For preparing it, best quality of *khand* is dissolved in boiling water till it matches with consistency of thick syrup. This is placed in a small porous earthen pot in which a cotton thread is suspended. When the sugar has crystallized, the vessel is inverted to drain off the excess mother liquor. It (vessel) is afterwards broken to secure the crystallized sugar in a lump form.

(h) *Sugar raw* - these are unrefined sugars under which come *bura*, or brown sugar *shakkar* or *masti meeja* or *talauncha* and other local brands.

Hulme and Sanghi (1918) have described in detail the indigenous methods for *gur* and sugar making. Indigenous factory—the *bel* for making *khandsari* contains five pans of the cheapest possible kind which are placed on a furnace made of mud in a thatched mud house. The pans vary in size. The juice is put into the largest pan and boiled and evaporated until its volume is reduced by 50 per cent. It is then ladled into the next pan and further evaporated and concentrated until it reaches the fifth pan where it is highly concentrated and made ready for crystallization. Thereafter, it is placed in a large earthen vessel about 60 cm high and 45 cm in diameter. After partial crystallization, stopper is taken out from the small hole in the bottom of the vessel for draining out the molasses. A river grass called *sewar (Vallisneria spiralis)* is placed on the top of the sugar to a thickness of 5 to 7.5 cm, this helps in removing the molasses on account of moisture and its power to retain heat. As a result bleaching of sugar occurs to a depth of 1.25 cm or even more.

The *sewar* is changed daily until the contents of the vessel have been treated. The sugar, taken off each day, is taken to *pata*, a level floor surrounded by a low wall. Here the sugar is dried in the sun and crushed by coolies by walking and twisting of their feet on it. This is called *swadeshi* sugar.

There is another indigenous process in which the *rab* is put into gunny bags. Six or seven bags are placed one upon the another; a coolie stands upon the top bag supported by a pole and sways forwards and backwards. As a result most of the molasses is separated from the sugar. The bags are then emptied into wooden bins and the sugar is treated with *sewar* grass as mentioned above. It is then sun dried and ground by coolies simply by walking on it with twisting movement of their feet. By this indigenous process the recovery of sugar is three tonnes per 100 t as against 10 t in tropical countries.

There is yet another process in which after concentrating the juice in five pans, the massecuite *rab* is put into *kalasis,* earthen vessels. Each vessel holds about 20 kg and when the material has crystallized, it is centrifuged to remove molasses. This molasses is again boiled and concentrated and a second sugar is made and the molasses is again spun off in the centrifuge. Molasses, thus obtained is rich in sucrose and it is sold for mixing with tobacco or making cattle feed. With this process, we recover sugar to the tune of four per cent.

In Gurdaspur, Barnes (1918) has described the manufacture of *khand* from *rab* as follows. The making of drained raw sugar (*khand*) is generally carried out by a regular trader. The process requires a good deal of supervision and few cultivators proceed further than making of the first crude sugar products - *gur* or *mal rab*. In making *khand,* the *mal rab* is emptied into large vats called '*kanchi*', which are made of wood and lined with matting, and have capacity of 3 to 150 qtl. of *rab*. At the bottom of the vat there are a number of small channels leading off to special reservoirs outside. On the flooring are placed pieces of wood and on these a reed mat is placed and over that a piece of coarse cloth the sides of which are stitched to the side mats on the vat is spread. After some time the molasses, known as *shira,* exudes through the cloth and matting at the bottom and passes in reservoirs and finally collected in earthen jars. After the *rab* has been in the vats for about 10 days and the mass has hardened sufficiently to bear a man's weight, it is worked up with an iron trowel so as to break all lumps, and then smoothed with a flat dish previously rubbed with *ghee*. The layers of *jhala (Potamogeton),* water plant, are placed on the top and after a few days the *jhala* is rolled - up and the white sugar at the top of the mass of *rab* is removed. Fresh *jhala* is put in its place and the old *jhala* is placed over it, so that the sugar is extracted. It takes 3 to 4 months to empty a vat in this way. This substance is called *khand* (sometimes also as *chini*).

From more superior ripe cane a more white product termed *shakkar* is obtained which is dry and more friable; this is usually sold in powder form. When boiling of juice is done for *shakkar,* the charge of juice is slightly reduced and the concentration carried a little further than when making *gur.*

The process of *rab* making differs from that of *gur* in the clarification of juice. After *sukhlai* or *suli (Euphorbia roylena)* is added, the contents of the pan are removed into large earthen jars and stored in a semi-solid, semi-liquid state until required for *khand* making. The proportion of juice to *rab* is 5 : 1 and of juice to *gur* is 6 : 1.

The composition of different sugar products is given in Table 8.1.

Kakavi (liquid gur)

It is prepared from the juice extracted from well matured canes. The juice is boiled in an iron pan on open fire to a striking temperature of 106°C. It is cooled, filtered and filled in sterile glass bottles. After filling the bottles, 0.04 per cent citric acid is added to reduce the formation of crystals due to increase in the levels of glucose or fructose in *kakavi* and also to impart characteristic colour. The mouth of each bottle is provided with an inner cap to prevent direct contact of *kakavi* with air. An outer screw cap is also provided. This liquid sugar is very popular in Maharashtra (Wandre *et al.,* 1991).

Technological Improvements in Gur and Khandsari Making

The *gur* or *khandsari* industry is mainly dependent on the introduction of improved bullock driven *kolhus* replacing the out-moted *kolhus* employed in the villages. Besides *kolhus,* improved juice boiling pans and furnaces have been introduced and propagated by the Khadi and Village Industries Commission (KVIC) with a view to ensuring better quality of products and higher returns to artisans. In order to take advantage of the economy of large scale operations, community *gur* and *khandsari* units have been sponsored by the KVIC since 1962-63 permitting thereby the use of power driven crushers, centrifuges and crystallisers.

Table 8.1: Composition of Various Indigenous Sugars

Description of the sugar product	Sucrose per cent	Glucose per cent	Specific gravity of solution	Total soluble matter	Moisture	Soluble organic and insoluble impurities
#*Gur* from *Dhalu* cane at Gurdaspur	73.21	13.34	1.0330	88.05	11.95	1.05
Shakar from *Dhalu* cane at Harchowal	75.60	9.29	1.0370	89.65	10.35	4.76
Rab (Shira)	41.02	20.68	1.0330	82.74	17.26	12.04
#*Khand* from *Dhalu* and *Katha* in Sri Govindpur growm in Beas Bet	91.81	1.30	1.0380	94.90	5.10	1.79
Khand from *Katha* in Harchowal	92.21	1.80	1.0378	94.40	5.60	0.39
Bura made in Quadian	81.04	6.45	1.0354	88.60	11.40	1.11
Misri made from *Khand*#	90.61	2.12	1.0375	93.60	6.40	0.87
Kuza misri	96.90	0.13	1.0389	97.00	3.00	nil

#Gur is very varibale in composition.
Source: Barnes, 1918.

In *khandsari* production, two processes, namely, liming and sulphitation were introduced in 1966 for ensuring better quality and higher recovery. With the introduction of hydraulic crushers coupled with liming and sulphitation process of juice clarification, it has been possible to obtain *khandsari* sugar recovery ranging between 7–7.5 per cent as against a recovery of 3.5 per cent recorded in the traditional process. It is noteworthy that even in the organised sector, the average recovery of sugar on cane basis has generally not gone beyond 9.5 to 10 per cent in spite of spectacular developments in the organised sugar mill sector in India. This reflects a creditable record insofar as recovery of *khandsari* sugar on cane basis is concerned.

Unlike *khandsari*, *gur* has certain inherent limitations as to the possibility of adopting chemical clarification of cane juice for obtaining refined quality product. In the manufacture of *gur*, the sugarcane crushers, namely, bullock or power driven based on the Indian Council of Agricultural Research, New Delhi and National Sugar Institute, Kanpur design and which are manufactured by the Government of India undertaking — The Nahan Foundry. This later became M/S Nahan Foundry Ltd.,—a Government of Himachal Pradesh enterprise. Crushers with specifications of the Nahan Foundry ones are now being manufactured by many other manufacturers and are being adopted presently. With regard to the design of furnaces and pans, those based on research work carried by the Indian Council of Agricultural Research, New Delhi and Planning, Action and Research Institute, Lucknow, have been adopted and recommended. The introduction of moulds of specified weight and shape, presentable packing and grading of *gur* under prescribed standards may help in convenient trading and availability of quality product to the consumers at a reasonable price.

Training in Gur and Khandsari Making

The advance and refresher courses of three and six months duration each in *gur* and *khandsari* industry are conducted departmentally by the *Gur* and *Khandsari* Training Centre, Kanwali, Dehra Dun (Uttar Pradesh). Besides, artisan training of one-month duration is also conducted by the State KVI Boards at their efficient production units (Anon., 1990).

Microstructure of *Gur*

Microstructure and chemical composition of *gur* are important from following view points: (a) to characterise the quality *gur* and (b) to determine which of the structural/chemical component in response to particular cultural practice, ambient conditions or the use of a particular clarificant is responsible for good quality *gur* and its subsequent deterioration during storage.

At the Regional Agricultural Research Station, Anakapalle (Andhra Pradesh), T. Varahalu studied microstructure and chemical composition of *gur* from 1934 to 1939. Varahalu (1937) reported that good quality *gur* sets quickly at higher temperatures, while the opposite is true for bad quality *gur.* The latter contained substances which cause high viscosities, retard the process of juice boiling and proved resistance to the formation of nuclei around which growth of sucrose crystals takes place. Thus, the degree of resistance, offered for free crystallisation of sucrose by the impurities present in the sugarcane juice, appears to be the key factor. Such a resistance was relatively more in the poor quality *gur* (Varahalu, 1938b).

When crystallisation takes place in a relatively shorter time at high temperatures, the crystals grow well and most of the enclosed fluid is extruded out. At high temperatures, the liquid would be less viscous and highly mobile. It is easily driven out centrifugally in the *gur* mould. Such a process occurs in good quality *gur* (Varahalu, 1937).

The good quality *gur* has a more open texture, contains relatively small quantity of water in equilibrium at relative humidity (RH) below 100 per cent and possesses relatively greater strength. Water in the *gur* is present either in 'free' or 'bound' form; the latter being held by the forces of absorption. It was also suggested that the moisture content of *gur* in equilibrium between 50-60 per cent RH could serve as a reliable index to evaluate *gur* quality (Varahalu, 1938a). Further studies indicated that good quality *gur,* though absorbed moisture under high humid conditions (at times even more than the poor quality *gur*), also parted with moisture in a similar fashion when dry conditions prevailed. Poor quality *gur,* however, absorbed moisture irreversibly (Narasimham *et al.,* 1965). Varahalu (1938b) also proved that apparent hardness of certain poor quality *gur* amounted to mechanical rigidity maintained by fluid film pressures and it was analogous to a kind of thixotrophy displayed by weak gels.

Later on Varahalu (1939) examined microtome sections of *gur* and correlated the properties of different types of *gur* with their structure and external features. In the light of this study, *gur* may be considered as polydispersed system wherein fractions of various particle sizes are distributed. It represents transitional forms varying from highly colloidal disperse systems through coarse disperse phases to the strictly rigid solids. The good quality *gur* has the rigid solid state with minimum internal energy content and the bad one represents several transition systems with greater, but varying energy contents.

Relationship Between Chemical Composition and Structure of Gur

Canes obtained from water-logged or swamp conditions give poor quality *gur* because the juice contains high organic non-sugars, high total and non-protein nitrogen, high colloids and pectins and a low amount of phosphorus compared to the juice yielding better quality *gur* (Mohan Rao *et al.,* 1951). This is also true when poor quality *gur* is obtained from crop fertilised with heavy nitrogen (Sanyal, 1926; Varahalu, 1937; Mohan Rao *et al.,* 1951). Such harmful effects of nitrogen manuring could, however, be avoided if it is supplemented with phosphatic fertilisers (Anon., 1959). Good quality of *gur* is associated with a low N/P ratio and this could be ensured either by phosphate manuring or the use of superphosphate in clarification (Tembhekar, 1954).

Farooque and Srivastava (1954) advocated a new physico-chemical concept of the hygroscopic nature of *gur.* They observed that *gur* colloids absorb and exchange cations like a clay colloidal system. The presence of sodium in the exchange complex was associated with a higher degree of hygroscopicity in *gur* as compared to colloids with other *cations* like potassium, magnesium and calcium. Calcium ions appeared to improve the keeping quality, while sodium affected it adversely. Same authors also reported that the reaction of residual tannins and iron present in the *gur* account for 50 per cent of the colour developed during storage (Farooque and Srivastava, 1952).

Suryaprakasha Rao and Mohan Rao (1959) presented a simple relationship between physico-chemical composition and structure of *gur* for evaluating its keeping quality and performance of various packaging materials for storage of *gur*.

Equilibrium Humidity

Relative Humidity at which a *gur* sample neither gains nor loses moisture at the ambient atmosphere is called equilibrium humidity. A lower value of this for a gur sample will imply that it begins to absorb moisture at rather a low RH of the ambient atmosphere. The higher the equilibrium humidity, the lesser would be the absorption of moisture. The work done at Anakapalle indicated that equilibrium humidity of good quality gur (Grade I) was in the range of 67.5 per cent relative to 66 per cent and 59 per cent of medium (Grade II) and poor (Grade III) quality gur, respectively. In other words, better quality gur (Grade I and II) would not begin absorbing moisture from the ambient atmosphere till the RH approaches a value of 66 per cent [Anon., (Anakapalle, India), 1973].

Dissolution Studies

Cubes of *gur* of various grades having the same surface area take different time intervals to dissolve in water when gently stirred. Its refraction may be monitored by a refractometer at successive 5 min intervals. Studies at Anakapalle indicated that *gur* of high quality took 55-56 min and that of inferior quality 35-40 min to reach the same stage of dissolution (in terms of refractometer value) [Anon., (Anakapalle, India), 1973].

REFERENCES

Anonymous (1959). Sugarcane research in India: 1929-1954—A review. *Indian J. Sugarcane Res. and Devpt.* 3 (pt. II) : 109-125.

Anonymous (1989). *Gur* manufacture by using sugar factory equipments. *S.I.S.S.T.A. Sugar J.*, 1(4): M14-M15.

Anonymous, (Anakapalle, India) (1973). Sixty Years of Agricultural Research at Anakapalle, SRS Anakapalle, A.P.A.U., Hyderabad, pp. 178.

Anonymous (1990). Glimpses of Cane Gur and Khandsari Industries under KVIC, Directorate of *Gur* and *Khandsari* Industry, Khadi and Village Industries Commission, 'Gramodya', Mumbai, India, pp. 46.

Barber, C.A. (1919). Progress of the sugarcane industry in India during the years 1916 and 1917. *Bull. No.* 83, Agricultural Research Institute, Pusa, Bihar, Supdt. Govt. Printing, India, Calcutta, pp. 45.

Barnes, J.H. (1918). Sugar and the sugarcane in the Gurdaspur District, *Technical Bulletin* No. 69, Agricultural Research Institute, Pusa, Bihar, Superintendent, Govt. Printing, India, Calcutta, pp. 99.

Farooque, M. and S.P. Srivastava (1952). (*c.f.* Anon., 1959).

Farooque, M. and S.P. Srivastava (1954). A new physico-chemical concept of hygroscopic property of *gur*. *Proc. Bien. Conf. S.R.D.W.I.* 2 : 602-607.

Hulme, W. and R.P. Sanghi (1918). The improvement of the indigenous methods of *gur* and sugar making in the United Provinces, *Technical Bulletin No.* 82, Agricultural Research Institute, Pusa, Bihar, Superintendent, Govt. Printing, India, Calcutta, pp. 22.

Mohan Rao, N.V., R.L. Narsimham and D. Anji Reddy (1951). Studies in quality of jaggery. *Proc. Bien. Conf. S.R.D.W.I.* 1 (2) : 143-151.

Narasimham, R.L., P.V. Suryaprakasha Rao, and T. Mallikarjuna Rao (1965). Some laboratory studies on physico-chemical aspects of *gur* storage. *Proc. Bien. Conf. S.R.D.W.I.* 5 : 484-492.

Pandit, S.N. and K. Singh (1983). Note on surfactant for *jaggery* making. *Maharashtra Sugar* 8 (3) : 31-34.

Sanyal (1926). (*c.f.* Anon., 1959).

Suryaprakasha Rao, P.V. and N.V. Mohan Rao (1959). Studies on the storage life of jaggery—Note on a quick and simple method of assessing the comparative keeping quality of jaggeries and evaluating the performance of various packaging materials. *Indian J. Sugarcane Res. and Dev.*, 3(2): 156-158.

Tembhekar, V.V. (1954). Effect of mineral ingredients present in sugarcane juice on the quality of *gur*. *Proc. Annl. Conv. D.S.T.A.* 11(1): 161-163.

Varahalu, T. (1935). Physico-chemical studies in sugarcane jaggery. *Madras Agril. J.*, 23(10): 389-393.

Varahalu, T. (1937). Studies in sugarcane jaggery. III. Emperical test for the end point of the boiling in jaggery manufacture. *Madras Agril. J.*, 25(12): 377-382.

Varahalu, T. (1938a). Studies in sugarcane jaggery. IV. Some properties of jaggery in relation to moisture. *Madras Agril. J.*, 26(5): 55-63.

Varahalu, T. (1938b). Studies in sugarcane jaggery. V. Macrostructure of jaggery. *Madras Agril. J.* 26(5): 394-398.

Varahalu, T. (1939). Studies in sugarcane jaggery. VI. Microstructure of jaggery. *Proc. Annl. Conv. S.T.A.I.* 8: 1-20.

Wandre, S.S., A.G. Shinde and A.B. Hasabnis (1991). Studies on keeping quality of *kakavi* (liquid jaggery) prepared at various striking temperatures and effect of citric acid on crystallization and colour, presented at *National Seminar on Increasing Productivity, Quality, Storability of Jaggery (Gur)*, held at the Indian Institute of Sugarcane Research, Lucknow, organised by the Indian Institute of Sugarcane Research, Lucknow and Indian Society of Agricultural Engineers, New Delhi, Nov, 30–Dec., 1991.

9

QUALITATIVE GRADING OF *GUR*

Quality and chemical composition of *gur* produced in different parts of the country vary widely due to differences in the sugarcane varieties used, ecological conditions, cultural practices and manufacturing techniques adopted. Quality and chemical composition are very important in the marketing of *gur*. Farmers bring their produce in the market in lots and it is purchased by the merchants. Visual observations such as colour, hardness, and texture are the common indicators used to judge the quality of *gur*. Flavour and hardness are judged by eating and pressing a small lump of *gur* between the two fingers or by scratching its surface with a knife. In the market different qualities of *gur* are exhibited for sales in lots based on merchant's ideas on quality. In fact superior quality *gur* of one market may be inferior to even the average one of another. A precise grading of *gur* is thus an inevitable need and more so for export purposes.

Criteria used for Grading *Gur*

A number of physical and chemical factors are important in *gur* grading. Amongst physical characteristics, colour, hardness, texture are important, while content of sucrose, reducing sugars, dirt, vegetable matter, taste, flavour and other impurities, ash and moisture are important chemical factors. Significance of these for rating quality of *gur* is discussed in the following:

Physical Characteristics

Colour : Amongst the physical factors affecting *gur* quality, colour is the most important one. For eating purposes light coloured *gur* is generally preferred.

Hardness : Hardness of *gur* depends on its moisture content. Hard *gur* generally contains low moisture and is preferred for storage.

Texture : Texture of *gur* may be crystalline or amorphous. It depends on the crystal content and size distributed in a lump of *gur*. It is a measure of purity of the product. A high crystal content coupled with crystalline texture indicates high percentage of sucrose. Crystalline condition is judged by rubbing the sample between the fingers. Crystalline *gur* is preferred for eating, while amorphous is used in confectionary.

Chemical Characteristics

Taste : *Gur* is sweet to taste. Sweetness is due to the presence of various sugars (sucrose, glucose and fructose). Fructose is the sweetest among the sugars present. Sweetness of *gur* depends on the total sugar content and also on the proportion of different sugars. Sometimes *gur* is saltish in taste. This is due to the high concentration of mineral salts. *Gur* of some of the varieties is slightly bitter in taste due to excessive caramelisation during boiling or deterioration during storage.

Flavour : Flavour is important for local trade and preference for a particular flavour depends on the food habit of a particular community.

Sucrose : Amongst all the chemical factors, it is the most important as it is the largest and chief sweetening component of *gur*.

Reducing sugars: Fructose and glucose are the main reducing sugars present in *gur*. Though they add to sweetness, they are hygroscopic in nature and adversely affect the keeping quality of *gur*.

Dirt and insolubles : Dirt and insolubles include bagasse, bagacillo and soil particles and the presence of these indicates defects in the manufacture, besides they adversely affect *gur* quality.

Ash : Inorganic constituent in cane juice and those added during *gur* manufacture constitute ash. Some of the ash constituents (Na and Cl) adversely affect keeping quality of *gur*. When present in excess, they also affect its taste.

Moisture: Large amount of moisture impairs the keeping quality of *gur*.

Schemes Proposed for *Gur* Grading

Keeping the above-mentioned quality factors in view, the following systems of *gur* grading have been proposed :

Agricultural Marketing Advisor's Scheme

The system adopted for grading *gur* under Agmark (1943) is based on the analyses of two sets of samples, the test sample and the check sample.

Table 9.1: Agmark System of *Gur* Grading

Grade	Speical characteristic (colour not darker than)	Texture	General characteristics of *gur* for grading purposes
Extra special	Creamy or light golden	Rawedar (crystalline or danedar (amorphous)	The *gur* should be prepared from the clarified juice of sugarcane Molasses shall not be used
Speical	Golden	-do-	Shall be prepared in the form of shape (solid lump) which shall be of firm consistency not sticky
A I	Dark golden	-do-	Shall reasonably be free from extraneous matter such as bagasse, dirt and other impurities and combined total of which shall be less than 1 % in weight
A II	Light brown	-do-	
B	Brown	-do-	Shall be dried to such an extent as reasonably as to maintain its colour Shall bear no superficial shoot or mould and Shall be sweet to taste, shall not possess a sour, salty or with other objectionable matter

Table 9.2: *Gur* Grading Chart

Quality	Grade	Particular of quality	Marks alloted
Sucrose (%)	1	80	40
	2	70	30
	3	60	20
	4	50	10
	Total no. of grades 4		Total marks allotted 40
Colour	1	*Kesaria*	20
	2	*Untia*	15
	3	*Pyazi*	10
	4	*Alsia*	6
	5	*Kathai*	4
	6	*Lohia*	3
	7	*Hathia*	1
	Total no. of grades 7		Total marks allotted 20
Texture	1	Granular	15
	2	Semi-granular	8
	3	Non-crystalline	2
	1		
	Total no. grades 3		Total marks allotted 15
Taste	1	Very sweet	10
	2	Sweet	8
	3	Bitterish	6
	4	Saltish	2
	5	Sweet saline	2
	Total no. of grades 5		Total marks allotted 10
Moisture (%)	1	1 to 3	10
	2	3 to 6	6
	3	6 to 8	2
	4	8 to 10	2
	Total no. of grades 4		Total marks allotted 10
Dirt (%)	1	Below 1	5
	2	1 to 2	3
	3	2 to 3	2
	Total no. of grades 3		Total marks allotted 5

Source: Khare, 1939.

In the scheme, five grades of *gur*, (1) Extra special, (2) Special, (3) A I, (4) A II and (5) B, are specified depending on the variations in colour, taste, consistency and presence of extraneous matter. This system of grading is used only for a lump of *gur*. The details of the Agmark System of grading are given in Table 9.1.

Earlier Khare (1939) had proposed grading of *gur* as shown in Table 9.2.

The *gur* is then divided into five grades on the basis of total marks obtained, as shown below:

Marks obtained	Designated grade
80 and above	Extra superior
70 to 80	Superior
60 to 70	Medium
50 to 60	Ordinary
40 to 50	Inferior

Gupta and Balyn (1973) classified *gur* into four grades on the basis of quality attributes, on visual observations of colour and texture, as given in Table 9.3.

Table 9.3: Grading of *Gur* based on Visual Colour and Texture

Grade designation	Physical character	
	Colour	Texture
Speical	Creamy of light golden	Granular
A 1	Dark golden	Granular or smooth
A-2	Light brown	-do-
A 3	Brown	-do-

*Source:*Gupta and Balyn, 1973.

Bharadwaj *et al.* (1977) prepared *gur* from different cane varieties, viz., Co 443, , Co 1148, Co 1336, Co 6425, Co 6425, CoS 510, CoS 673, CoS 718, Co J 64, BO 70; and on the basis of physico-chemical properties suggested four grades of *gur* as Special, A1, A2 and B Table 9.4.

On the basis of four grades proposed above, *gur* prepared from cane varieties, viz., Co 1336, Co J 64, Co S 611, Co S 718 and BO 70 was of special grade. *Gur* from varieties Co 443, Co 1148, Co 6425, Co S 510 and Co S 673 was placed in A1 grade.

Gur Grading in Uttar Pradesh

In Uttar Pradesh, a major *gur* producing State in India, attempts have been made to grade the *gur* produced. Kirtikar *et al.* (1962) collected *gur* samples from Ghaziabad, Meerut, Muzaffarnagar, Saharanpur, Aligarh, Moradabad, Shahjahanpur, Kanpur, Faizabad, Varanasi and Gorakhpur and graded them into four groups based on physical and chemical characteristics (Table 9.5).

Table 9.4: Physico-chemical Attributes for *Gur* Grades

Physio-chemical attribute	Grades of *gur*			
	Speical	A 1	A 2	B
Texture	Granular	Granular	Amorphous	Amorphous
Status	Hard	Hard	Soft	Soft/melts
Pol (%)	Above 70	66-70	60-65	Below 60
Purity (%) Coefficient	Above 80	70-80	60-70	Below 60
Colour*	Above 100	100-150	150-200	Above 200
Net rendement	Above 55	45-55	40-55	Below 40

* Colour of half normal solution of *gur* was determined using a Klett Summerson colorimeter (660 nm filter).
Source: Bharadwaj *et al.*, 1977.

Table 9.5: Grades of *Gur* Produced in U.P.

Characteristic	Grades			
	I	II	III	IV
Colour	Light brown	Brown	Reddish brown	Dark red
Sweetness	Good	Good	Fair	Ordinary
Flavour	Good	Fair	Ordinary	Ordinary
Sucrose (%)	74.9	73.5	69.3	61.5
Invert sugar (%)	8.65	10.55	13.88	17.28
Ash (%)	2.58	2.44	2.51	2.66
Purity (%)	82.2	81.2	76.3	67.5
pH	5.5	5.5	5.6	5.6
Lead number	19.30	20.46	20.41	22.48
Net rendement value	57.2	54.4	46.6	34.9
Grade on net rendement value	II	II	IV	IV

Source: Kirtikar *et al.*, 1962.

Table 9.6: Chart of the Proposed Grades of *Gur* Produced in Different Zones of Uttar Pradesh

Characteristic	Grades of *gur*					
	Extra special	Special	A	B	C	D
Colour (units)*	50	51-100	101-150	151-200	201-250	Above 250
Sucrose (%)	Above 85	80-84	75-79	70-74	65-69	Below 65
Reducing sugars (%)	Below 5	6-10	11-15	16-20	21-25	Above 25
Moisture (%)	Below 3	3-5	5.1-8	8.1-12	12.1-20	Above 20
Insoluble residue (%)	Below 1	1.1-2	2.1-3	3.1-4	4.1-5	Above 5
Ash (%)	Below 1	1.1-2	2.1-3	3.1-4	4.1-5	Above 5

*Colour was measured in 13% *gur* solution by Klett Summerson colorimeter (660 nm filter).
Source: Dubey and Lal, 1989.

Dubey and Lal (1989) collected *gur* produced in different zones of Uttar Pradesh, viz., Eastern, Central, *Terai*, Western and Bundelkhand and categorised them into six grades based on physical and chemical properties. The grades were : (1) Extra special, (2) Special, (3) 'A' grade, (4) 'B' grade, (5) 'C' grade and (6) 'D' grade (Table 9.6).

The different *gur* grades should have the following physical characteristics:

Extra special: Crystalline, tasty, hard in nature and possessing extremely pleasant flavour.

Special : Crystalline, tasty, hard in nature possessing very pleasant flavour.

'A' grade : Crystalline, tasty, hard in nature, possessing pleasant flavour.

'B' grade : Less crystalline, tasty, hard in texture, less in flavour.

'C' grade : Small crystals but taste *at par* with 'B' grade.

'D' grade : Dark brownish colour, no crystals, poor in taste, soft in nature and possessing a bad smell.

Thus, on the basis of above gradation, it is suggested that only *gur* of Extra special, Special, A, B and C grades should be used for eating purposes. For storage purposes, *gur* of Extra special and Special grades is better. *Gur* of A and B grade can also be stored but only for limited period. *Gur* of these grades should be stored during monsoon period. The *gur* of grade C and D should not be stored at all.

Gur Grading in Maharashtra

Gur produced in Maharashtra has been categorised into three grades based on physical and chemical properties (Table 9.7) by Dangre (1963). *Gur* belonging to grade I and II is generally grouped in one lot because little variation occurs in the chemical composition. It is designated as of good quality. Grade III may be designated as of poor quality.

Gur Grading in Bihar

Sugarcane Research Station, Pusa, Bihar has suggested a system of *gur* grading in which five factors are used. These include total sugars, visual colour (as judged by caramel standards), texture, hardness and taste. Grading is done by allotting as many marks as the percentage of total sugars in *gur* and deducting marks for other quality factors as per equation:

Grade = S–Vc–Vtr–Vh–Vt, where

S = per cent total sugars in *gur*,
Vc = value for colour,
Vtr = value for texture,
Vt = value for taste and
Vh = Value for hardness.

Table 9.7: *Gur* Grading in Maharashtra

Characteristic	Grade		
	I	II	III
Sucrose (%)	78 - 85	75 - 78	70 - 75
Invert sugar (%)	5 - 8	8 - 10	10 - 15
Moisture (%)	3 - 5	5 - 7	7 - 10
Ash (%)	below 2	2 - 3	3 - 4
Organic matter (%)	3 -5	5 - 7	7 - 10
Colour (with 3.25% *gur* solution)			
Red filter - log T	2.00 - 2.05	2.05 - 2.30	2.10 - 2.20
Green filter - log T	2.10 - 2.15	2 .15 - 2.30	2.30 - 2.40
Blue filter - log T	2.15 - 2.30	2.30 - 2.40	2.40 - 2.50
Visual colour	Golden yellow	Dull yellow	Brown
Texture (sugar grain size between)	D and E	E and F	F

Source: Dangre, 1963.

The marks deducted for the various quality factors are given in Table 9.8.

Gur is graded into three grades, based on the marks, as follows:

Grade 1 - above 60
Grade 2 - 45 to 60
Grade 3 - below 45

In this scheme the marks deducted for colour and texture are based on market value and the relationship of hardness to moisture. The maximum moisture in *gur* is 10 per cent. Hence 10 marks are deducted for the softest *gur.* The basis for deducting marks for other categories of taste have been arbitrarily considered.

Table 9.8: Marks to be Deducted for Various Quality Factors

Quality factor	Particulars of quality	Marks to be deducted
Visual colour	Below 9 % caramel	0
(as compared with	9 - 16 % caramel	10
standards containing)	16 - 28 % caramel	15
	above 28 % caramel	20
Hardness	Above 704	0
(g cm^{-2})	7.04 - 704	5
	Below 7.04	10
Taste	Sweet	0
	Saltish - sweet	5
	Bitterish - sweet	10
Texture	Crystalline	0
	Semi - crystalline	5
	Non-crystalline	10

Source: Anonymous, (Lucknow, India), (1958-62). *Consolidated Final Report of the Scheme for Studies on Physical and Chemical Criteria of Gur Leading to the Formulation of the Basis of Gur Grades*, Indian Institute of Sugarcane Research, Lucknow, pp. 41.

Gur Grading in Andhra Pradesh

In Andhra Pradesh, commercial grading of *gur* (*jaggery*) is done only on the external physical characteristics, like, colour and consistency. Chemical composition was incorporated in its grading later. *Jaggery* samples of I, II and III grades were collected from some important regional centres in Andhra Pradesh and analysed for sucrose and total non-sugars. The colour and pore space were also determined. Initial physical and chemical analyses of all grades of *jaggery* were done and later periodical analyses from time to June to November were carried out for moisture, sucrose, glucose and colour on N/2 solution of *jaggery* in water.

The percentage of sucrose and non-sugars, among the chemical characteristics, and colour on N/2 *jaggery* solution, in the physical attributes, appear to be the most important in affecting *jaggery* quality. These have been taken as criteria for grading *jaggery* (Table 9.9).

Table 9.9: Grading *Jaggery* in Andhra Pradesh

Grade of jaggery	Limits for classsification		
	Sucrose per cent	Total non-sugar per cent	Colour * reading of N/2 *jaggery* solution
I	81-84 and above	< 5	< 135
II	78-81	5-8	135-225
III	77-75	> 8	>225

* 65 g *jaggery* in 500 ml of distilled water.
Source: Anonymous (Anakapalle, India), 1989.

Gur Grading in Tamil Nadu

In Tamil Nadu, Chockalingam (1985) graded *gur* into five categories (Table 9.10).

Table 9.10: Proposed Grades of *Gur* Prouced in Tamil Nadu

Characteristic	Grade				
	Extra speical	Speical	AI	AII	B
Colour	Creamy or light goldern	Goldern	Dark golden	Light brown	Brown
Texture	Granular	Granular	Granular	Granular	Granular

Source: Chockalingam, 1985.

The standard *gur* should not contain less than 65 per cent sucrose. It should have less than five per cent reducing sugars and three per cent insoluble residues. The sulphated ash in standard *gur* should not exceed 3-4 per cent (Chokalingam, 1985).

Gur Standard

As per Bureau of Indian Standards. (IS 12923, 1990),

requirements for sugarcane *gur* are as follows:

Characteristic	Requirement for	
	Grade 1	Grade 2
Sucrose (on dry basis) % by mass, minimum	80	70
Reducing sugars (on dry basis), by mass, maximum	10	20
Moisture %, by mass, maximum	5	7
Water insolubles (on dry basis), % by mass, maximum	1.5	2
Sulphated ash (on dry basis), % by mass, maximum	3.5	5
SO_2 (on dry basis), ppm, maximum	50	50
Ash insouble in dilute hydrochloric acid (on dry basis) % by mass, maximum	0.3	0.3

Note: Total sugar (expressed as overt sugar) on dry basis shall be 90 per cent for both the Grade 1 and Grade 2 cane *gur*

Sources: Anonymous (1990). Bureau of Indian Standards, IS 12923, 1990; (Clause 4.3 and 6.1), New Delhi, India.

Export Potential

The demand of *gur* in Africa, parts of Europe, Gulf countries and in Latin American countries is gradually increasing. But they demand it in uniform shape and size prepared hygenically with attractive packing. Thus, *gur* can also be a good foreign exchange earner.

In order to have a uniform quality of *gur,* Central Purchasing Centres for export of *gur* are the need of the hour. Keeping this in view Ghosh and Agrawal (1983), have proposed five grades of *gur* on the basis of physico-chemical properties after analysing *gur* samples collected from different *gur* producing areas in India.

The details of five grades of *gur* are given in Table 9.11.

Besides special characters given in Table 9.9, in general, *gur* should be sweet to taste with a palatable flavour. Normally, the ash content of *gur* is not considered for grading, though large quantities of ash make the *gur* saltish. The upper limit of ash content in *gur* for (grading purposes) should not exceed six per cent. Use of hydros in *gur* manufacture brightens the colour but is injurious to health. As per Prevention of Food Adulteration Rules of Government of India, the sulphur di-oxide content should not exceed 70 ppm.

In local market *gur* is usually purchased by visual assessment of physico-chemical properties. The methods suggested by Agricultural Marketing Advisor (1943) and Khare (1939) are not precise and dependable. For this reason rapid methods for the assessment of colour in solid phase, total sugars, *gur* hardness and moisture have been evolved.

Table 9.11: *Gur* Grading for Central Purchasing Centres

Quality factor	Tolerance limits for				
	Grade 1	Grade 2	Grade 3	Grade 4	Grade 5
Colour * (N/2 solution)	Below 125	125-174	175-274	275-499	Above 500
Texture Crystal (%)	Above 60	Above 60	30-60	30-60	Below-30
Moisture (%)	Below 5	Below 5	5-7.4	5-7.4	7.5-10
Sucrose Crystal (%)	Above 80	Above 80	70-80	70-80	60-70
Reducing sugars (%)	Below 5	Below 5	5-9.9	5-9.9	10-15
Insoluble impurities (%)	Below 1	Below 1	1-1.9	2-2.9	3-4
Sulphur dioxide (ppm)	Below 70	Below 70	Below 70	Below 70	Below 70

*Colour was determined 13% solution of *gur* using by Klett Summerson's Colorimeter (660 nm filter)

Source: Ghosh and Agarawal, 1983.

Table 9.12: Colour of proposed *gur* in solid phase

Colour square in Dictionary of Colour*	Grade proposed
10 E 4	Grade 1
11 F 5	Grade 2
12 F 8, 13 G 8	Grade 3
14 E 10, 7 C 11, 8 H 11	Grade 4
15 C 10, 16 C 10, 48 J 8	Grade 5

* The numerical preceeding the letter of the alphabet in the colour designation above refers to the serial number of plate in the Dictionary of Colour.

Rapid (Portable Kit) Method for the assessment of Colour in Solid Phase, Total Sugars, Hardness and Moisture

Colour in Solid Phase

A simple method has been evolved at the Indian Institute of Sugarcane Research, Lucknow for quick determination of colour standard for grading of *gur* in solid phase (Anon., (GGS), 1955-59), by comparing *gur* colour with the colour squares given in the *Dictionary of Colour* (Maer and Paul, 1950) (Table 9.12).

Farooque and Shrivastava (1964a), have prepared colour standards for *gur* grading by mixing 2, 4 and 8 g of caramel with 50g of crushed sugar and rubbed with two drops of glycerine and a few crystals of mercuric chloride so as to give it a texture and enhance its keeping quality.

The visual depiction of colour corresponding to the numbers is given in Fig. 19.1.

The final colour standards were prepared by mixing 0.5, 1.25, 2.25, 3.0, 6.23, 7.50, 10.0 and 11.5 g lots of caramel with 24.4, 23.75, 22.75, 18.75, 17.50, 15.00 and 13.75 g of crushed white sugar, respectively and rubbing it with two drops of glycerine and a few crystals of mercuric chloride. These were transferred to test tubes, corked and sealed. The standard thus prepared contained 2, 5, 7, 12, 25, 30, 40 and 45 per cent caramel in sugar. The results were expressed in terms of per cent caramel in white sugar.

Moisture per cent Gur

Farooque and Shrivastava (1960) developed a rapid colorimetric method for the estimation of moisture in *gur*. The method consists of shaking 2 g of *gur* in a test tube with 10 ml of an alcoholic solution of reagent cobalt chloride. A colour chart has been prepared representing the colour developed with 10 ml of colour developing reagent shaken with 2 g of *gur* containing 2.5, 5.0, 7.5, 10.0, 12.5 and 15.0 per cent moisture and shown in colour plate.

By comparing with colour chart, the moisture content can be calculated (Fig. 19.2).

(*Reagent* : One gram anhydrous cobalt chloride is dissolved in 2 ml of concentrated HCl and diluted with absolute alcohol to 500 ml in a measuring flask.)

PERCENT MOISTURE IN GUR

(WHEN 2G GUR + 10.CC REAGENT TAKEN)

MOISTURE	
BELOW	1%
	2·5 %
	5·0 %
	7·5 %
	10·0 %
	12·5 %
	15·0 %
	17·5 %

PERCENT REDUCING SUGAR WHEN VOLUME OF N/2 GUR SOLUTION TAKEN

1CC	2CC
	1·9 %
—	2·9 %
—	3·8 %
9·6 %	4·8 %
11·5 %	5·7 %
13·5 %	6·7 %
15·4 %	7·7 %
17·3 %	8·6 %

Reducing Sugars in Gur

Farooque and Ghoshal (1960) devised a rapid colorimetric method for the estimation of reducing sugars in *gur*, as described hereafter:

2 ml of N/2 *gur* solution (13 g/100 ml water) is taken in a test tube of 2.5 cm diameter, 5 ml of reagent No. 1 is added, heated on a spirit lamp to just boiling and cooled. Thereafter, 5 ml of reagent No. 2 and 3 ml of reagent No. 3 are added. The test tube is corked, shaken and the developed colour is compared with the chart which directly gives the percentage of reducing sugars in *gur*.

In case where the entire quantity of ferricyanide is consumed and no colour is developed on the addition of reagents No. 2 and 3, the experiment is repeated with 1 ml of the *gur* solution, the colour is read in the second column of the colour chart (Fig. 19.3).

Reagents:

(1) 8.25 g of potassium ferricyanide and 10.5 g anhydrous sodium carbonate are dissolved in water and made up to 250 ml. The solution is kept in a dark coloured bottle for 2-3 days before use.

(2) 12.5 g potassium iodide, 25.0 g zinc sulphate and 125 g sodium chloride are dissolved in water and made to 500 ml.

(3) 5 ml of glacial acetic acid is diluted to 50 ml.

Total Sugars in Gur

Under Agricultural Produce (Grading and Marketing Act, 1937), specifications of standard grades have been prescribed for sugarcane *gur*. According to these specifications, the product specified "Hard *Gur*" should contain more than 80 per cent sucrose by weight. In the absence of any quick rapid (portable kit) method, Farooque and Shrivastava (1964b) have devised a rapid method for the estimation of total sugars in *gur*.

Rapid method consists of a preparation of N/2 *gur* solution (65 g *gur* dissolved in 500 ml of water). About 20 ml of N/2 *gur* solution is taken for the estimation of total sugars. First the refractometer brix of above solution is recorded. Then the same is treated with phosphate-free activated charcoal, shaken and filtered after a minute or two. The first few ml of filtrate is again transferred to the filter paper and subsequent portion collected. Finally, the refractometer brix of the filtrate is recorded. Total sugars in *gur* is calculated by the formula:

$$\text{Total sugars \% } Gur = 7.7\ [a - 1.77\ (a - b)]\ \text{sp. gr.},$$

where

a = Refractometer brix of unclarified N/2 *gur* solution
b = Refractometer brix of clarified N/2 *gur* solution
sp. gr. = Specific gravity of unclarified N/2 *gur* solution.

For the convenience of field workers a ready reckoner in terms of formula has also been prepared from the brix of clarified N/2 *gur* solution and the difference between brix of unclarified and the clarified N/2 *gur* solution, the total sugars can be read (Table 9.13).

Table 9.13: Ready Reckoner for the Estimation of Total Sugars

Brix of unclarified N/2 *gur* solution	Differences between clarified and unclaried N/2 solution					
	0.4	0.6	0.8	1.0	1.2	1.4
10.6	79.4	76.6	73.7	70.9	68.0	65.2
10.8	81.1	78.2	75.4	72.5	69.7	66.9
11.0	82.2	79.9	77.2	74.2	71.4	68.5
11.2	81.4	81.6	78.7	75.9	73.0	70.2
11.4	86.1	83.3	80.4	77.6	74.7	71.9
11.6	87.8	85.0	82.1	79.3	76.4	73.5
11.8	89.5	86.6	83.8	80.9	78.1	75.2
12.0	91.2	88.3	95.5	82.6	79.8	76.9
12.2	92.9	90.0	87.2	84.3	81.5	78.6
12.4	94.6	91.6	88.9	86.0	83.2	80.3

Source: Farooque and Shrivastava, 1964b.

Texture (Crystal Content) of Gur

A rapid method for the determination of texture (crystal content) of *gur* was evolved at the Indian Institute of Sugarcane Research,

Lucknow (Anon., (GGS), 1961-62). Two gram of *gur* was weighed into a porcelain dish and rubbed with small quantity of rectified spirit saturated with sugar. The content was transferred to a 15 ml centrifuge tube using the sugar saturated spirit and volume made up to 15 ml. It was then centrifuged for 15 min at 2500 rpm. The volume of solid crystalline matter in centrifuge tube was read, halved and multiplied by 100 to give the percentage of crystals in *gur*. This method gave results comparable with the gravimetric method (Table 9.14).

Hardness of Gur

Farooque and Nazir (1960) have evolved a rapid method for the determination of *gur* hardness at Pusa, Bihar. For evaluating *gur* hardness, rind hardness apparatus designed by Khanna (1935) has been successfully used after certain modification. The design of the modified apparatus is given in Fig. 9.4.

Table 9.14: Crystal Content of *Gur* Estimated by the Rapid and Gravimetric Methods

Sample No.	Crystal content of *gur*	
	Rapid method	Gravimetric method
1	60.00	62.00
2	75.00	80.00
3	65.00	64.00
4	80.00	78.70
5	80.00	83.40
6	90.00	90.00

Source: Annl. Rept. Gur Grading Scheme, IISR, Lucknow, 1961-62.

Hardness is indicated in one pound of pressure required to press a needle of 0.35 cm diameter thickness through *gur* block up to a depth of 1 cm. The mean of four observations per block, two on the surface and two at the bottom as worked out at 97 per cent probability levels gives the hardness. Hardness of *gur* decreases with increase in moisture content.

An apparatus for the rapid determination of hardness of *gur* samples has been designed and fabricated at the Indian Institute of Sugarcane Reserach, Lucknow (Anon., (GGS),

1961-62). This instrument consists of a vertical brass tube 'A' approximately 2.5 cm in diameter and exactly 56 cm in height (Fig. 9.5). This tube is provided with two horizontal cross arms 'BB' in the middle of its length, carrying at their free ends two guide bushed 'CC' which are parallel to tube 'A'. The guide bushes slide on two vertical cylinder guides 'DD' fixed on a

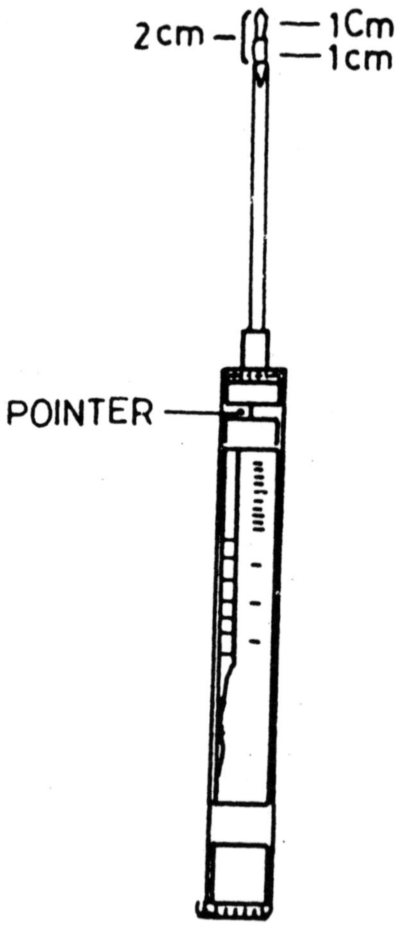

Fig. 9.4: Modified Hand Penetrometer apparatus for hardness (Farooque and Ahmad, 1960)

base plate 'E'. A needle 'F', 2.5 cm in diameter and 2.5 cm in length is fixed to the bottom of a cylindrical weight 'G' capable of sliding freely inside tube 'A'. Combined length of weight and the needle is 6 cm. The needle and the cylindrical weight together weigh exactly 100 g. Needle itself has a conical head (4 mm in diameter) with an apex angle of 60°. The weight with its needle is held in position inside tube 'A' with the top of

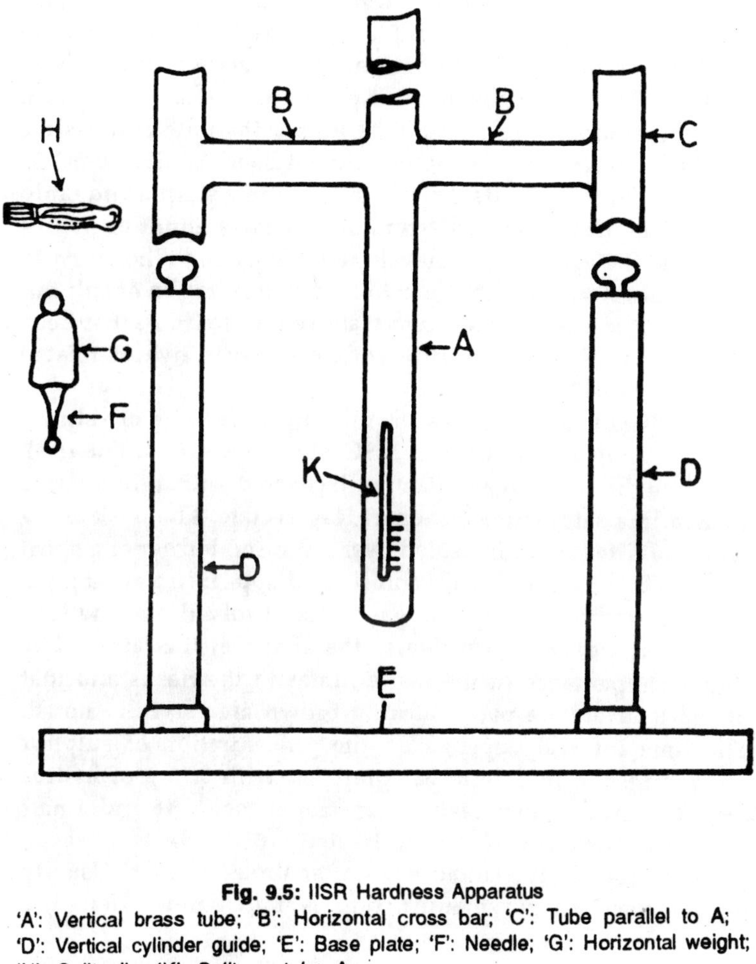

Fig. 9.5: IISR Hardness Apparatus
'A': Vertical brass tube; 'B': Horizontal cross bar; 'C': Tube parallel to A; 'D': Vertical cylinder guide; 'E': Base plate; 'F': Needle; 'G': Horizontal weight; 'H': Split clip; 'K': Split on tube A

the weight exactly on a level with the top of the tube by means of a split clip 'H'.

Gur hardness is determined in either top or bottom and middle of two *gur* blocks per sample. The reciprocal of the average of the four observations is taken as hardness of a *gur* sample.

Hydros Content in *Gur*

Sodium hydrosulphite, commonly known as hydros, is used as a bleaching agent at the time of *gur* manufacture. This improves the colour of *gur* but at the same time leaves behind certain amount of sulphur dioxide in the *gur* which increases with an increase in the amount of hydros used. The amount of sulphur dioxide should not be higher than the permissible limit of 70 ppm as per the Prevention of Food Adulteration Act of Government of India. Due to this reason a simple and rapid test for the detection and estimation of the amount of hydros added to the *gur* has been developed (Sastry and Ghosh, 1964). In the test, a scale for the detection of concentration of sulphur dioxide in relation to time in *gur*, above certain limits, has been set up to enable one to differentiate between hydros-treated and untreated *gur*.

To 20 g of *gur* taken in a 250 ml conical flask 3 g of sulphur free granulated zinc and 40 ml HCl (1 : 3) are added. The flask is thoroughly shaken and its mouth covered with a filter paper soaked in a saturated solution of lead acetate. The flask along with the filter paper is heated over a Bunsen burner or a spirit lamp. White fumes of hydrochloric acid appear in a short time. Besides, hydrogen sulphide gas is also evolved and the lead paper turns dark brown due to the action of the latter. The time of appearance of the white fumes in the flasks and that of the lead acetate paper turning brown are carefully noted. The time interval depends on the concentration of sulphur dioxide in *gur*. In case of *gur* prepared with 0.25 g of hydros per litre of cane juice lead acetate paper took one and a half minute to turn brown. *Gur* untreated with hydros, however, did not produce the colour even after three minutes. Usually hydros used in *gur* manufacture is much more than the recommended level.

REFERENCES

Anonymous (1943). Report of Marketing of Sugar in India and Burma by Agricultural Marketing Advisor, pp. 222-223.

Anonymous (GGS) (1955-59). *Annl. Rept. Gur Grading Scheme*, Indian Institute of Sugarcane Research, Lucknow, India.

Anonymous, (GGS) (1961-62). *Annl. Rept. Gur Grading Scheme*, Indian Institute ot Sugarcane Research, Lucknow, India.

Anonymous (1990). Bureau of Indian Standards, IS 12923, 1990; (Clause 4.3 and 6.1), New Delhi, India.

Bharadwaj, K.M., M. Singh and Rajendra Singh (1977). Significance of physico-chemical test for *gur* grading. *Indian J. Sugar Crops* 4(4) : 94-97.

Chockalingam, S. (1985). *Gur* manufacture in Tamil Nadu–A Review. *National seminar-cum-group discussion on jaggery manufacture and storage* held at the Indian Institute of Sugarcane Research, Lucknow (Dec., 18-19, 1985).

Dangre, A.J. (1963). Studies on *gur* qualities in Maharashtra. *Proc. Annl. Conv. S.T.A.I.*, 31 (pt. I): 105-113.

Dubey, V. and U. Lal (1989). *Gur* grading in Uttar Pradesh. *Indian Sugar*, 39(2): 101-106.

Farooque, M. and Nazir Ahmad, (1960). A simple method for determination of *gur* hardness. *Proc. All India Conf. Sugarcane Res. and Dev. Workers* 4: 410.

Farooque, M. and R.K. Ghosal (1960). A rapid method for estimation of reducing sugars in *gur*. *Proc. All India Conf. Sugarcane Res. and Dev. Workers*, 4: 421-424.

Farooque, M. and S.P. Shrivastava (1960). A rapid method for estimation of moisture in *gur*. *Proc. All India Conf. Sugarcane Res. and Dev. Workers*, 4: 482-484.

Farooque, M. and S.P. Shrivastava (1964a). Colour standard for *gur* grading. *Proc. All India Conf. Sugarcane Res. and Dev. Workers*, 5: 482-484.

Farooque, M. and S.P. Shrivastava (1964 b). Rapid method for the estimation of total sugars in *gur*. *Proc. All India Conf. Sugarcane Res. and Dev. Workers* 5: 492-494.

Ghosh, A.K. and M.P. Agrawal (1983). *Gur* grading based on physical and chemical constituents. *Maharashtra Sugar* 8(12) : 39-43.

Gupta, D.N. and S.V. Balyn (1973). Chemical criteria for grading of *gur*. *Sugar News* (Feb.): 20.

Khanna, K.L. (1935). An improved instrument for testing the rind hardness in sugarcane. *Agriculture and Livestock India*, 5(1): 19.

Khare, L.N. (1939). Grading factors and methods of grading *gur*. *Proc. Annl. Conv. S.T.A.I.* 8: 535-553.

Kirtikar, C.P. Gupta and M.K. Srivastava (1962). Criteria of *gur* grades in Uttar Pradesh. *Proc. Annl. Conv. S.T.A.I.* 30 (Pt. I) : 104-110.

Maerz, A. and M. Rea Paul (1950). *Dictionary of Colours*, McGraw-Hill Book Company, London, England.

Sastry, A.S. and A.K. Ghosh (1964). A rapid test for detecting the use of hydros in the manufacture of *gur*. *Indian J. Sugarcane Res. and Devpt.*, 8(4): 470.

10

GUR STORAGE

In India, about 40 lakh tonnes of *gur* is produced annually. The production of *gur* starts from October and continues up to May, depending on the location. As all the *gur* produced is not consumed soon after production, it is stored roughly for a period of 10-12 months. The major problem associated with *gur* storage is its ability to absorb moisture, particularly during the monsoon season. Often during storage due to high moisture the *gur* liquefies, making it useless for human consumption. On a conservative estimate 10 per cent of the total *gur* produced is damaged during monsoon resulting in a net loss of about Rs. 40 crore every year. To mitigate these losses different methods for *gur* storage have been developed. These can be divided into two broad categories; (i) small scale storage (by farmers) and (ii) large scale commercial storage.

Small Scale Storage of *Gur*

On small scale *gur* is stored by the farmers in the following ways:

(i) In a small earthen pot properly covered with lid. The lid is covered with a piece of clean muslin cloth (Uttar Pradesh and Punjab).

(ii) In air tight earthen pots which are kept hanging from the roof to protect them from ants (Assam).

(iii) A layer of trash and hessian cloth or bags is spread on the floor and *gur* blocks are arranged in a row over it (Maharashtra).

(iv) For short periods *gur* is stored in paddy straw (Tamil Nadu).

(v) In some areas, such as Visakhapathnam district of Andhra Pradesh, *gur* is allowed to settle in big capacity

(0.8 qtl) earthern pots. The mouth of each pot is covered with leaves and the pot is stored upside down on a nearly 15 cm thick layer of sand.

(vi) Earthern pots, being semi-pervious in nature, allow migration of water vapour both ways and to stop this these are painted from outside. *Gur* stored in such pots can be kept for a longer time as compared to unpainted ones (Baboo and Shukla, 1987).

(vii) In Madhya Pradesh, farmers store *gur* in burnt mud pitcher. The capacity of the mud pitcher varies from 20-80 kg. If the size of the pitchers is small, they are stacked one over the other. The pitchers are kept in the corner of a room over a layer (10-15 cm) of straw/dried sugarcane tops. The pitcher is filled with *gur*, its mouth closed by the lid and sealed with mud mortar to make it air-tight. But, in fact, it is never cent per cent air-tight because the wall of the mud pitcher is semipervious and water vapour can migrate both ways. As a result, the storage induces changes both in colour and consistency of the *gur*. Because of these changes the farmer quickly disposes of his product (Chouksey, 1985).

(viii) Agrawal and Ghosh (1984a) have advocated the use of *gur* storage pit for small farmers. The detailed design of the pit is as follows:

A pit measuring one metre square at the top and 0.75 metre at the bottom and one metre deep is dug in the mud floor of the *varandah* of a rural house. The bottom and the sides of the pit are coated with a 2 cm thick mud plaster and allowed to dry slowly without developing cracks. When the plaster dries up completely, a 5 cm thick coat of bitumen and sand plaster is applied. For this, hot bitumen is first applied on the mud plaster surface with a brush and dry sand is then dusted liberally over the bitumen. Sand dusting is continued until bitumen stops flowing out.

Blocks of *gur* of four different qualities, wrapped in a newspaper to prevent direct contact with the bitumen, are stacked inside the pit just before the onset of monsoon. The pit is then covered with a flat lid made up of wooden planks. Over this lid a 2 cm thick layer of mud plaster is applied extending well beyond the edges of the wooden lid on all the four sides. This plaster is allowed to dry slowly. On the top of the mud plaster, a 5 mm thick coat of bitumen and sand plaster is applied in the same fashion as advocated for inside surface of the pit. The pit may be opened in October and the *gur* consumed or sold in the market. Nearly 7 qtl of *gur* can be stored in one pit. It is more effective when the *gur* storage pits are constructed in a covered space, e.g., *verandah* etc. It is possible to use same pit year after year at least for 5 consecutive years without incurring any further expenditure. The storage capacity of these pits, however, is flexible and may be kept according to need.

Analytical data of *gur* lots before and after storage in pits located in open and covered space are given in Table 10.1.

(ix) Baboo *et al.*, (1985) have suggested the use of *gur* drying-cum-storage bin of one qtl capacity for a small farmer having six family members (Fig. 10.1). The details of this bin are given below:

This metal bin is of galvanised iron sheet of 60 cm diameter and 90 cm height. Four holes of 2.5 cm diameter are made at equal peripheral distance from one another. The holes for air inlet are made at 5.0 cm height from bottom. A support is made, in the shape of a cross, of two wooden pieces of size 2.5 × 5.0 × 60 cm. The support is placed at the bottom of the bin in such a way the air inlet hole lies in between two consecutive corners of the support. A circular (60 cm dia.) G.I. sheet of 20 gauge is cut and then holes of 10 mm diameter are made in it. The holes are made at 25 mm radial distance at a peripheral spacing of 25 mm from each other. This perforated sheet is placed over the wooden support.

A lid of 20 gauge galvanised iron sheet is prepared for the bin. In the centre of the lid, a hole of 20 cm diameter is projected to a height of 15 cm and another small lid is made to cover the projected hole. The projected holes serve the following two purposes: viz. (i) it is used as support to keep the standing solar aspirator around it for drying *gur* and (ii) whenever one needs small quantity of *gur* during the monsoon season it can be taken out through these outlets.

Table 10.1: Changes in the Composition of *Gur* before and after Storage in Underground Pits

Lot No.	Constituent	Before storage	After storage in pits placed in	
			Open	Covered space
1	Per cent gain (+) or loss (–) in weight	100	99.6	99.4
	Moisture (%)	100	127.6	91.3
	Sucrose (%)	100	102.3	96.1
	Reducing sugars (%)	100	100.9	109.8
	Colour in solution	100	122.0	109.3
	pH	100	100.0	100.0
2	Per cent gain (+) or loss (–) in weight	100	100.0	104.7
	Moisture (%)	100	126.7	102.2
	Sucrose (%)	100	98.7	99.4
	Reducing sugars (%)	100	101.8	108.4
	Colour in solution	100	195.1	100.1
	pH	100	96.3	100.0
3	Per cent gain (+) or loss (–) in weight	100	97.8	104.2
	Moisture (%)	100	124.6	118.0
	Sucrose (%)	100	101.7	99.5
	Reducing sugars (%)	100	99.6	110.2
	Colour in solution	100	119.8	106.5
	pH	100	98.1	99.1
4	Per cent gain (+) or loss (–) in weight	100	99.6	101.8
	Moisture (%)	100	131.8	122.9
	Sucrose (%)	100	98.7	99.5
	Reducing sugars (%)	100	114.9	111.5
	Colour in solution	100	124.0	107.7
	pH	100	98.1	100.0

Gur of two shapes, viz., *balti* and *laddu* are dried in the bin during April. Data on *gur* drying in the bin

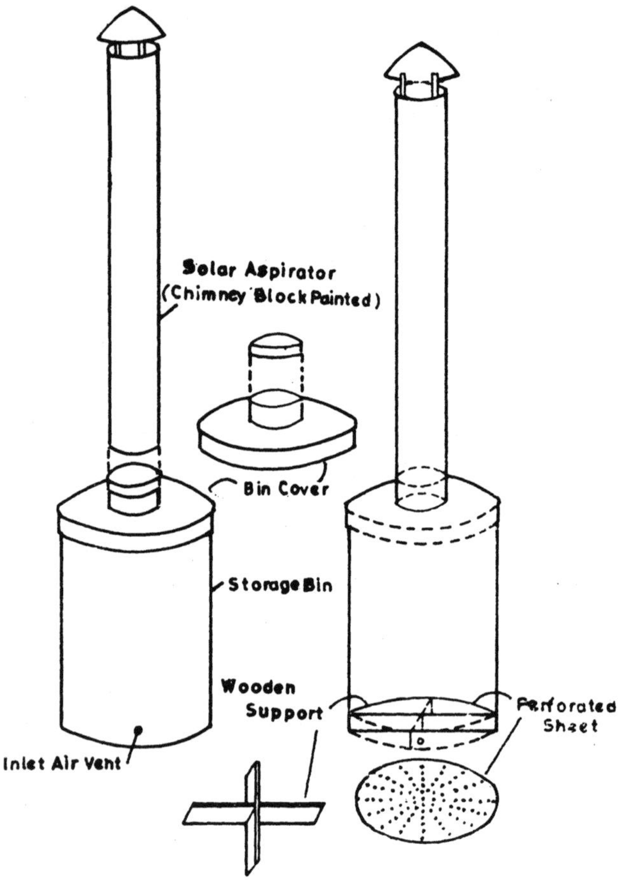

Fig. 10.1: *Gur* drying-cum-storage bin (Baboo *et al.* 1985)

are presented in table 10.2a, 10.2b. *Gur* is dried to a moisture level below 5.50 per cent in 6 to 10 days during hot summer days.

Gur dried in the bin can be effectively stored during monsoon season as it absorbs only up to 10 per cent moisture.

Parthasarathy and Lakshmikantham (1950) have described some methods of *gur* storage in vogue in the southern part of India.

In Tiruchirapalle, Madurai and Coimbatore (Tamil Nadu), small quantities of *gur* for domestic consumption are stored in earthen pots. The mouth remains covered with cane trash and sealed with cow dung and red soil. Merchants store *gur* in gunny bags inlaid with palm leaf mats. Each bag contains 140 lb (64 kg approx.), and bags are kept separately on floor and not one over the other. Storage in bulk is also done by spreading a layer of cane trash, 15-20 cm thick, on the *pucca* floor of the room on which *gur* cubes are spread evenly to a height of about 23 to 30 cm. Three layers of *gur* cubes (alternate with trash layers) are spread for storing *gur*. Paddy husk or ground nut shell in place of cane trash is also used if *gur* is kept in gunny bags.

Table 10.2a: Drying of *Gur* in Drying-cum-Storage Bin

Shape of *gur*	Gur weight (kg) on the day							Initial moisture content % (db)	Final moisture content % (db)	Moisture loss % (db)
	Initial	2	4	6	8	10	12			
Balti	100	98.40	97.10	96.10	95.30	94.75	94.75	9.50	5.25	4.25
Laddu	100	98.10	96.40	95.20	95.20	95.20	95.20	8.20	4.80	3.40

Weather during the period (April 16-28, 1985).
Daily mean temperature = 31.2°C.
Daily mean relative humidity = 36.34%.

In south Karnataka, solid *gur* in the form of moulds weighing 3-6 kg is packed in trash and stored in a place having access to smoke from kitchen or the hearth. It has been seen that *gur* could be preserved well under such conditions.

Table 10.2b: Moisture Absorption by *Gur* in Drying-cum-Storage Bin

Shape of *gur*	Initial moisture content % (db)	Moisture content (%db) of *gur* after storage			Net gain in moisture % (db)
		One month (Jul. 20)	Two months (Aug. 20)	Three months (Sept. 20)	
Balti	5.00	5.35	5.60	5.85	0.85
Laddu	4.50	4.90	5.25	5.50	1.00

Atmospheric conditions of the storage room:
Monthly mean temperature °C: July = 30.85, August = 31.89 and September=31.58
Monthly mean relative humidity (%) : July = 75.5, August = 72.14 and September = 72.34.
Source: Baboo *et al.* (1985).

Storage of *Gur* on Commercial Scale

(i) *Gur* is stored in closed rooms after spreading a 7.5 cm thick layer of trash/bagasse/*bhoosa* on the floor and then spreading a mat over it. The *gur bhelies* are stacked over these (western Uttar Pradesh). Sometimes, *gur* is packed in gunny bags before stacking (Andhra Pradesh and some parts of western Uttar Pradesh).

(ii) *Gur* is stored on raised wooden or bamboo platforms about 90 cm above the ground, with *bhoosa* lining on top and bottom and covered with hessian bags on all sides (eastern Uttar Pradesh, Bihar, Bengal and Orissa).

(iii) *Gur* is stored in empty kerosene oil tins. The mouth of each tin is closed with a small piece of cloth. A layer of dry leaves, 5.6-7.5 cm thick is spread over the tins.

(iv) In Kolhapur, coastal region of Maharashtra, *gur* is stored in heaps of cotton seed in godown having a few windows. Bamboo mats are spread over the floor and then a layer of cotton seed about 10 cm thick is spread over it. *Gur* blocks, wrapped in hessian cloth, are stacked side by side in one layer. Above them a 7.5 cm thick layer of cotton seed is again spread. In this way *gur* is stacked in three tiers. A moisture absorbing substance, such as, quicklime or calcium chloride is kept in the room to ensure relatively drier conditions.

(v) In the Eastern Godavari district of Andhra Pradesh, *gur* is stored in specially constructed godown by generating smoke. This methods is relatively more effective than other methods in preserving *gur*. The godown in this case is a *pucca* building made of bricks and cement. The floor comprises of a 6 cm concrete bed overlaid with a 30 cm layer of compacted soil. The inner walls are plastered with cement, while outside ones are coated with mud. The roof is made of corrugated cement sheets. The godown has eight rooms, in two rows of four each. Each room has four storage racks, the first one being 17.5 cm from the floor and others at nearly 15 cm distance from each other. The racks are made of bamboos, closely placed and nailed on to supporting wooden beams. The dimension of each room is 5.40 m × 3.00 m and is provided with one door of 1.95 m × 1.05 m and six ventilators 90 cm × 30 cm in size. Only the lower three tiers have two windows each. The top-most has none. A cement channel, 22.5 cm wide × 22.5 cm deep, is provided all-around the godown and filled with water to protect the *gur* from ants.

Gur slabs weighing 28.26 kg each, wrapped in sugar cane trash, are placed in palmyra leaf baskets which are stacked on the floor up to four layers.

The paddy husk is used for producing smoke. It is spread evenly in a 10 cm layer. For a floor space of 16.5 sq. metres, seven bags of paddy husk (120 kg) are sufficient for each smoking. Ventilators are closed and the husk is burnt by means of cow dung cakes placed in four corners of the room and the door is closed. Each smoking takes about four days and it is done once in a month during summer and thrice during monsoon to lower the moisture inside the godown. Due to the heat evolved during these 'smokings', the temperature of the godown as also *gur* is increased and concomitantly the relative humidity inside the godown goes down below the one prevailing outside.

Due to smoking the rise in temperature is in the range of 5.5° - 11.0° C and then reached within two days after burning of paddy husk. Thereafter, it takes 6-7 days to come to the normal temperature. During this period relative humidity inside the godown ranges from 65 per cent to 75 per cent as against 85 per cent – 95 per cent outside.

The ventilators are kept open during dry weather. Smoking in summer is done to reduce the excess moisture absorbed by *gur* during storage as also the moisture from the palmyra leaf baskets. The capacity of each room is nearly 1100 packages of *gur*. In this system of storage, little or no change occurs in sucrose content of *gur*. There is, however, slight reduction in *gur* weight and its colour darkens after some time. The loss in weight in *gur*, under this method of storage ranges from 1 to 2 per cent only.

(vi) At Dankaur, in western Uttar Pradesh, generally houses are built in two floors, ground and the first floor. The family of merchant lives on the ground floor. *Gur* is stored on the first floor in rooms, leaving a central open space and with an inside *varandha* about 90 cm wide. The rooms may be of any length and width but invariably about 3.30 m in height. The roof is made of wooden planks overlaid with bricks and cement. Bamboo mats are placed on the floor and the *gur bhelies* are stored in 4 or 5 layers. These are then covered with hessian bags. A room of about 54.0 m × 54.0 m × 36.0 m holds about 25.9 t of *gur*. The *gur* is kept during hot summer months in the storage rooms. The windows, facing east and west, are kept open at this time in order to drive out the moisture from the *gur*. When the rainy season starts they are closed.

About 100 godowns of the type mentioned above in Dankaur village alone store nearly 3600 mt of *gur*. The *gur* is generally purchased and stored in March or April and is taken out and sold from August onwards.

(vii) At Muzaffarnagar, Central Warehousing Corporation has built godowns for large scale storage of *gur*. The capacity of each godown is 1250 mt. Godowns are fitted with exhaust fans on the western and eastern sides. About 40 kg of *gur* is packed in hessian bags. These are stored over steel racks and crates erected inside the godown. In each rack, bags are stacked in 3 to 4 layers. Exhaust fans fitted in western side draw air inside the godown while the ones on the eastern side blow it out of the godown. These exhaust fans are operated during the day time throughout the summer months (April to June). Before the onset of monsoon, the ventilators, fan spaces and doors are sealed with polyethylene sheets.

Comparing relative humidity and temperature at Rajamundry (Andhra Pradesh), Kolhapur (Maharashtra), Muzaffarnagar and Dankaur (western Uttar Pradesh), Sastry and Ghosh (1965) established that the most important factor in storage (at Muzaffarnagar and Dankaur) is low relative humidity 48 to 49 per cent in the summer month (June) (Table 10.3).

(viii) At Sikandarabad, a small township near Dankaur, the godowns are constructed on the same principle except that the storing capacity of each godown is about 59.2 mt.

(ix) At Gaziabad, there are big godowns divided into 18 rooms, in two rows of 9 each with a central passage, which is about 2.4 cm wide. Each compartment is subdivided into 4 racks 3 m × 1.2 m × 1.5 m. On the floor, a 10 cm layer of *bhoosa* is spread and is covered with hessian bags to prevent rise of moisture from the floor. *Gur* is stored over this in 5 layers. In other racks no *bhoosa* is spread and there is no matting. There are two windows of 1.2 m × 0.69 m one above the other on the outer side of each compartment. These are kept open in summer and electric fans in the central passage ensure proper aeration.

Table 10.3: Mean Temperature and Relative Humidity in Godowns at Different Trade Centres in India

Period (month)	Location							
	Rajamundry		Kolhapur		Muzzafarnagar		Dankaur	
	T	RH	T	RH	T	RH	T	RH
June	35.7	71.04	30.1	67.49	33.5	48.05	35.6	49.01
July	27.2	86.61	26.7	87.33	33.2	68.30	28.2	70.87
August	28.3	73.80	27.2	83.22	31.3	75.50	25.7	83.50
Sept.	27.1	78.51	26.7	87.57	31.5	71.75	28.4	73.29

Source: Sastry and Ghosh, 1965.

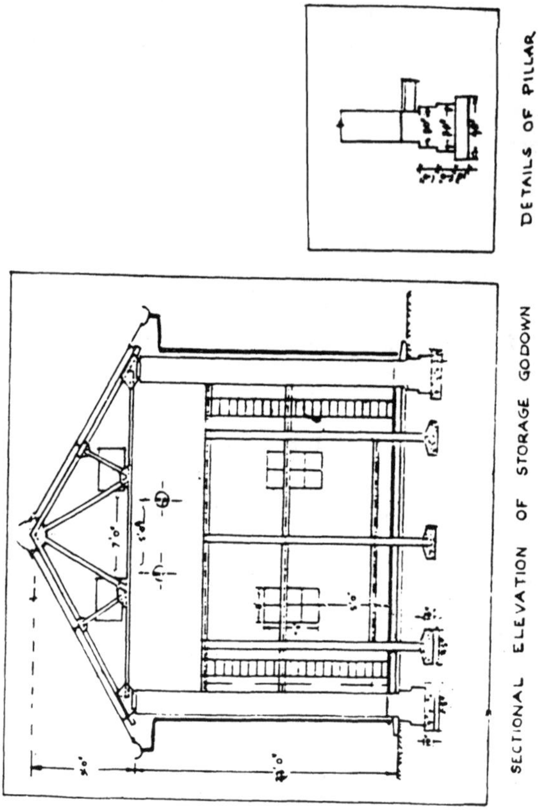

Fig. 10.2: Details of Constructional Features of Godown (Verma *et al.*, 1958)

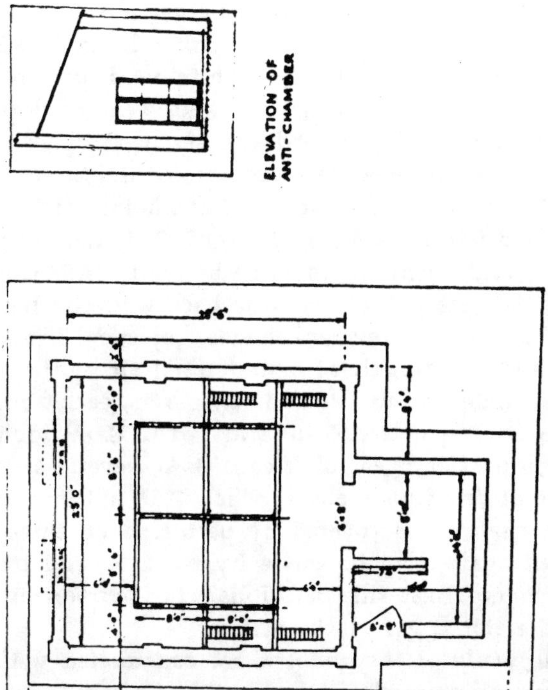

Fig. 10.3: Detailed Plane of the Basic Unit of the Godown (Verma *et al.*, 1958)

(x) A godown of dimensions 25 ft (7.63 m) length, 28 ft (8.54 m) and 22 ft (6.71 m) high, area 700 sq ft (65.16 m) with an ante-chamber having an entrance of 5 ft (1.53 m) × 8 ft (2.44 m) × 10 in. (25.4 cm) thick baffle wall in between entries to guard against direct entry of humid air was designed by Verma *et al.* (1958) (Fig. 10.2 and Fig. 10.3). The ante-chamber has a separate galvanised iron sheet roofing fixed on 2.5 inch (6.35 cm) × 2.5 inch (6.35 cm) × 5/16 inch (0.079 cm) angle iron frame. Four ventilators 2 ft 6 in. (1.07 m) with an 1/4 inch (0.64 cm) thick perplex sheet pane fixed and barricade of closely spaced 5/8 in (1.59 cm) mild steel bars provide for light, two on the front and two on the back wall at a height of 22 ft (6.71 m). Two windows 3 ft (0.92 cm) × 5 ft (1.53 m) glazed inside and barricaded outside by 5/8 in (1.59 cm) mild steel bars are provided on the back wall at a height of 6 ft (1.83 m). Window rain shades project out 18 in. (45.70 cm) and are capable of being closed air tight.

The godown consists of 12 chambers, each about 8 ft × 8.5 ft (2.44 m × 2.59 m) and 7 ft (2.14 m) high. The platform of each chamber consists of several stretchers made of dried bamboos 3 inches (7.62 cm) in diameter. Each chamber is covered by partitions on three sides which again may be made by tying several bamboo stretchers. These side partitions act as supporting walls for the stored *gur* blocks or bags.

The walls of the godown are constructed with clay mortar and painted with lime *surkhi.* Cement is generally avoided. No rain water from the roof should find an access into the godown. Seepage of moisture through the foundation wall can be avoided by applying moisture proof material on the foundation of the godown at the floor level i.e. at the level from which the walls are raised. The roof of the godown is constructed with corrugated galvanized iron sheets and not cement asbestos sheets. A gutter of ample capacity to drain off the rain water from the roof through a number of drain pipes is provided. Inside and outside walls of the godown are coated with asphalt. The inside coating is thicker, up to a level of 6 ft (1.83 m) to 8 ft (2.44 m).

Double doors are provided in the godown with a view to preventing outside air from entering the godown whenever it is opened for storing/taking out *gur*. Only the outside door is to be opened and closed again. In region where high humidity conditions exist and the electricity is available, dehumidifiers can be installed inside the godown at each corner. These can absorb moisture from the inside atmosphere and the absorbed water can be drained out. Alternately several trays containing silica gel or quick lime can be kept on the ground floor. In place of humidifier two 16 inch (40.64 cm) sweep exhaust fans can be fitted on the front wall at a height of 17 ft (5.19 m) with air-tight covers and having semi circular rain shades.

In this godown *gur* has been effectively stored in the relative humidity range of 70 to 80 per cent.

(xi) At Gudiatham (Tamil Nadu), an improved *gur* godown of an overall dimension 13.5 m × 5.6 m × 4.42 m having a capacity of 200 mt has been designed by Subramaniam and Vaidyanathan (1964). The godown has a single storeyed central hall and single room of double storey. The floor of the godown is lined with 2.54 cm thick asphalt layer and the roof is constructed with 15.24 cm concrete slab over which brick tiles are laid giving a slope for the drainage of rain water. Four exhaust fans are provided, two in each room of the first floor. To prevent moisture ingress during rainy season, doors and windows are sealed with rubber lining.

In this godown *gur* can be stacked up to 15 m layer on the floor in heaps or bags.

(xii) At the Indian Institute of Sugarcane Research, Lucknow (Uttar Pradesh), Panje and Menon (1964) designed a *gur* storage godown which enables the drying of *gur* in the pre-monsoon period but can be sealed during the monsoon. The design aims at admitting the dry air during the hot weather, ensuring a high degree of sealing against ingress of moisture and providing a fair amount of thermal insulation at relatively low cost. The basic unit of this design is capable of storing 50 t *gur*, the smallest economic size. The storage capacity can, however, be

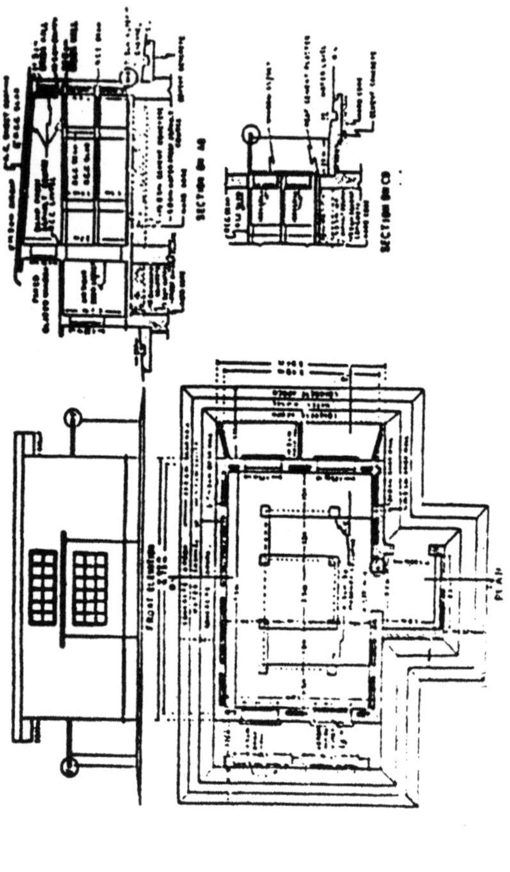

Fig. 10.4: Detailed Plan on the Basic Unit of the Godown for Drying and Storage (Panje and Menon, 1964)

increased by duplicating the unit either along its length or width, or quadruplicating by eliminating the middle walls.

The basic unit (Fig. 10.4) consists of a room 7.5 m long, 5m wide and 3.7 m high at the crown end of the roof and 3.4 m high at the eaves end. The walls of this room are of hollow cavity type, constructed of solid brick masonary in 1 : 6 cement mortar. The outer one end of the double walls is 23 cm thick and the inner 11.5 cm thick with an air passage of 11.5 cm in-between them for thermal insulation. The foundation is of solid brick masonary in 1 : 6 cement mortar on base concrete 1 : 5 : 10. The roof is also double and consists of an ACC roof over a RCC slab with air passage of 11.5 cm in-between them. This roof is laid with a slope of 1 in 15 from North to South to ensure rapid drainage of rain water. Two tiers of RCC shelves 1.25 m wide are provided along the walls, leaving a passage 1.25 m wide between the rows of shelves to permit easy handling of the product. The shelves are supported on 41 cm square brick columns spaced 1.25 cm apart.

To prevent seepage of moisture, the floor is lined with a 5 cm thick waterproof asphalt cover over 15 cm of 1 : 2 : 4 cement concrete and the internal surface of the walls, ceiling and shelves are also finished with a damp-proof asphalt course.

The entrance into the godown is through an ante-chamber 2.5 m × 2 m and 2.5 m high. The door leading from the ante-chamber into the godown (Fig. 10.5) is, when closed, bolted down against the surface of the walls and made air-tight by a rubber lining on all four sides between the door frame and the wall surface.

The godown has to be located along the direction of the dry winds in the locality so that these winds can be utilised to dry out the *gur* whenever necessary. The use of draft fans or other machinery for ventilation can thus be avoided. *Gur* when received for storage is stacked on the shelves, the space in front of the window opening being left empty until the last so as to ensure effective ventilation inside the godown for as long a period as possible, i.e. up to the time the humid easterly winds start. A step ladder can be used for loading the top shelves. The door between the ante-chamber and the godown

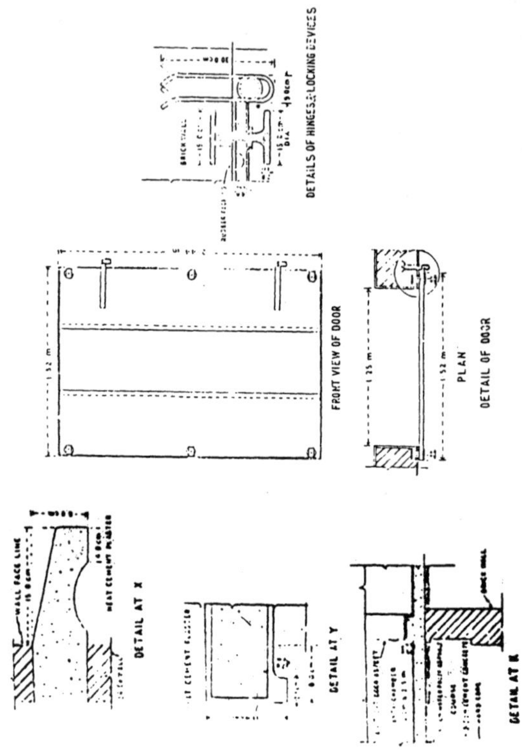

Fig. 10.5: Details of Constructional Features of Godown (Panje and Menon, 1964)

is kept bolted in position at all times excepting when *gur* has to be taken out of the godown. During the spring and early summer, the windows are left fully open at night and the early hours of the day when the humidity is below 50 per cent and the atmospheric temperature is low. Care should be taken to close the windows and doors whenever there are any sudden thunder showers and hailstorms during the hot weather or if at any time the wind direction changes to easterly and humidity increases above 50 per cent or the outside temperature rises beyond 30°C. When nights are dry, the windows can be kept open at night and closed at about 4 A.M. so as to keep the godown cool as well as dry. As soon as the moist winds start blowing before the break of monsoon the windows are bolted tight against the walls, to be opened again only after the monsoon. Thus, the keeping quality of *gur* improves by a careful manipulation of the godown ventilation. This can be greatly facilitated by installing a large size thermo-hygrometer on the outside of the front door for periodic monitoring of the ambient conditions inside the godown.

For removing *gur* from the godown during the monsoon, care has to be taken each time that the front and the back doors of the ante-chamber are not left open at the same time. The windows should not be opened on any account until the weather becomes dry again. The water channel on the apron should not be allowed to dry.

In some areas, *gur* is stored in a large scale in well ventilated rooms keeping alternate layers of bagasse and *gur* (Parthasarathy and Lakshmikantham, 1950). In Hindupur (Andhra Pradesh) *gur* pallets (20-30 g) are heaped on the cemented floor to a height of about 1.2 m and turned over ones in a two-month period. In Srikakulum (Andhra Pradesh) *gur* is put in small earthen pots which are placed inverted one over the other. The floor of the room is sprinkled with sand. In Anakapalle (Andhra Pradesh) 4 to 5 layers of *gur* buckets wrapped in hessian cloth are stacked in a room.

The work done at the Sugarcane Experiment Station, Cuddalore (Tamil Nadu) revealed that for small scale storage alkathene lined gunny bags or air-tight drums are ideal containers. For large scale storage, however, air-tight asphalt floored scientific godowns appeared to be ideal (Shakunthala, 1981).

Storage of *gur* in Potato Cold Storage

Khanna and Chacravarti (1955) indicated that keeping quality of *gur* can be maintained in cold storage or in a refrigerator, as compared to all other methods available. On the basis of earlier finding, Chacravarti *et al.* (1969) evaluated the potato cold storage for preservation of *gur*. Potato cold storage has two chambers:

(a) The temperature of the main chamber varies from 15° to 30°C with relative humidity from 90 to 95 per cent.
(b) The temperature of the precooling chamber varies from 18° to 22° C with relative humidity from 55 to 60 per cent.

The results indicated that 25 per cent loss in weight accompanying liquefaction, putrefaction and all round deterioration can be prevented by storing *gur* under controlled conditions prevailing in the cold storage and there was no change in weight and chemical characteristics.

Pre-cooling chamber of cold storage which remains vacant during potato storage can profitably be utilised for *gur* storage with hessian cloth packing (or even no packing). In cooling chamber *gur* can be kept after packing in polyethylene or hessian cloth coated with wax (of which former might be preferred).

Absorption of moisture and deterioration of *gur* samples depend on the relative humidity in the atmosphere than on a value where *gur* neither gains nor loses moisture and weight (equilibrium humidity, EH). Equilibrium humidity for *gur* lies between 55 and 65 per cent RH where loss in weight is minimum (Sastry and Ghosh, 1964). At low RH of 30 to 40 per cent, appreciable moisture loss occurs leading to a loss in pressure bearing capacity. It may be mentioned here that a reasonably good quality *gur* generally withstands a pressure of 0.5 lb/square inch (0.035 kg/cm^2) (Surya Prakash Rao and Mohan Rao, 1959).

On the basis of the above findings, Gundu Rao and Ramaiah (1961) have designed a godown to store 350 mt *gur*. The size of the godown is 30′ × 30′ × 30′ (9 m × 9 m × 9m).

They have also fixed some parameters for construction of the godown, namely, selection of site and general layout etc.

Gur Storage

Selection of Site

Godown should be constructed on a raised ground, the level of which should be higher than its surroundings. Water should not stagnate or enter into the godown. It should be constructed preferably near the *gur* producing centre and should be well connected with roads and uninterrupted electric supply.

General Layout of the Godown

The godown has 900 sq. ft (81 sq. m) floor area. It has three chambers built on R.C.C. floor with wooden frame 4′ × 27" (1.2 m × 67 cm or 8′ × 27" × 4.5" (2.4 m × 67 cm × 11.25 cm). Each chamber has a capacity to store 100 to 120 mt of *gur*.

The walls and floor of the godown are to be plastered with lime *surkhi* instead of sand cement mixture. Roof is made of R.C.C. slabs with 3" (7.5 cm) slope in 15 ft (4.5 m) from the middle of roof. Roof has to be laid out with a thick coating of asphalt and coarse gravel. A double door system is provided to prevent moisture ingress. It is better not to have any ventilator in the godown. If ventilators are to be provided they should be so constructed as to avoid moisture ingress.

Drains of the rain water have to be constructed all round the godown at least three feet (90 cm) away from the godown wall.

The godown requires air conditioning for the efficient storage of *gur*. A 30′ × 30′ size godown containing roughly 8 to 10 thousand cu ft of free air requires 1.5 ton to 2 ton air-conditioning unit operated at 15°C and 50 per cent RH.

Effect of Inorganic Constituents on the Keeping Quality of *Gur*

Quality of the *gur* prepared from sugarcane varieties Co J 79, Co 1148 and Co J 77 deteriorates during storage. Deterioration is due to high amount of calcium, magnesium and chloride in *gur*. Phosphorus decreases deterioration by 9 to 10.6 per cent. It absorbs up to 6 per cent moisture and sucrose content decreases by 7 to 8.2 per cent whereas reducing sugars show an increase of 4.8 per cent. The increase in colour intensity, however, is maximum in the *gur* from Co J 64 and Co J 71 having the highest nitrogen content (Kapoor and Kanwar, 1990).

Effect of Pressure on the Storage and Keeping Quality of *Gur*

It is a common experience that when *gur* bags or lumps are stacked one over the other in tiers, *gur* in the lower tier deteriorates more rapidly than in the upper ones. This is due to the pressure exerted by the bags in upper layer. Situation worsens if relative humidity (RH) inside the godown is high. Besides these two factors, physical and chemical composition of *gur* also play an important role in its keeping quality. *Gur* stored in lower tiers could withstand a pressure of 35 g cm^{-2} at 72 per cent RH, while at 58 per cent RH the same *gur* could withstand thrice that pressure i.e. 105 g cm^{-2} (Agrawal and Ghosh, 1984b).

Deterioration of *Gur* in Storage

During storage of *gur*, considerable amount of sugar is lost as CO_2; the maximum being in the moisture range of 6-10 per cent. This indicates the importance of low moisture in the product stored. Critical maximum and minimum levels of moisture for the growth of microorganisms are 20 and 3 per cent, respectively (Anon., (Pusa, India), 1953-54).

During storage, particularly under high moisture, *gur* deteriorates rapidly by the activity of microorganisms which is purely a surface phenomenon. The surface layers (0-0.6 cm) exhibit maximum growth of microorganisms. Their population decreases rapidly up to depth of 2.0 cm and below this practically sterile condition prevails under actual storage conditions (Farooque, 1957). Twenty-two strains of yeast have been isolated from samples of *gur*, molasses and cane syrups by Ligange *et al.* (1982). The predominant one is *Saccharomyces cerevisiae*. Microorganisms associated with the deteriorated *gur* include twelve fungi, one bacterium, one *Actinomyces* and different species of the yeast (Jadhav *et al.,* 1984).

Singh *et al.* (1978) have reported that shape of *gur* is related to the rate of deterioration. *Gur* shapes displaying large exposed area are liable to deteriorate faster. Less deterioration has been observed in case of small ball shaped *gur* as compared to bigger ones. Lal and Sharma (1983) have observed that spherical or hemispherical shapes of *gur* are better for storage than the angular ones. It has also been observed that deterioration of *gur* first starts in the surface layers and the powdered *gur* deteriorates quickly. On the other hand, the semi-liquid *gur* stored in containers and covered with hessian cloth shows least change during storage. Baboo and Ghosh (1984) have observed the relative humidity above 62 per cent is conducive to moisture absorption. Moisture absorption and desorption appeared to be confined to surface layers (2.5 cm) of the *gur*.

Effect of Shape and Size of *Gur* in Storage

The shape and size of *gur* lumps vary appreciably from place to place. In different parts of India *gur* is available in various shapes like basket, small cube, spherical ball, etc. The weight of the individual lumps varies from 0.050 kg to 50 kg (Table 10.4).

Effect of Shape on Drying

Drying of *gur* is necessary for proper storage. It can be dried up to 60°C without any adverse effect on the quality. The drying of *gur* is a surface phenomenon and is generally restricted to 2.0 cm surface layer. *Gur* with initial moisture of five per cent when dried at 60°C showed a loss of 23.60, 15.38 and 11.49 g. kg^{-1} in case of *laddu, chaku* and *dhayya,* respectively (Baboo and Ghosh, 1985).

Packaging Materials for *Gur* Storage

Studies on packaging material for *gur* have been carried out at the Central Sugarcane Research Station, Pusa, Bihar. The materials tried included furnace ash, powdered charcoal and wheat *bhoosa.* Best preservation of *gur* was achieved where samples were kept in furnace ash. The loss in storage was about 12 per cent as against 34.3 per cent with *bhoosa*. Suitability of wrapping materials and containers for *gur* storage indicated that the following material are suitable. Paper coated with a paraffin wax or a mixture of paraffin wax and an oil in the proportion of 2 : 1; hessian bags painted with white paint; hessian bags coated with a preparation of linseed oil and glue; and *gur* stored in hessian bags and then embedding them in furnace ash.

Table 10.4: Shape and Size of *Gur* Manufactured in India

State	Shape of *gur*	Size / weight (kg)	Local name
Andhra Pradesh	Basket shape	10	*Dimma, laddoo* bricks
Assam	Earthen pot shape	3-20	—
Bihar	Lumps bigger size	20-40	*Chakki, laddoo* and cakes
Karnataka	Small cubes	—	—
Maharashtra	*Mattas*	20	*Bheli*
Orissa	Earthen pot shape	—	—
Punjab	Cake shape	5	*Pansera bheli*
	spherical	4	*Chausera bheli*
Tamil Nadu	Small tapering cubes small balls	0.01-0.02	Cakes
Uttar Pradesh	Semi spherical/lumps	0.05-0.10	*Laddu*
	Semi spherical/lumps	2-3	*Dhayya*
	Semi spherical/lumps	1-2	*Pari*
	Trapezoidal lump	4-5	*Chaukhunta*
	Semi spherical/lumps	5-6	*Pansera*
	Semi spherical/lumps	9-10	*Dhansera*
	Tapered cylindrical lump	1-20	*Balti*
	Trapeozoidal slab broken into irregular pieces	0.5-3	*Chaku*
	Small trapeozoidal slab	2.5-5.0	*Khurpa pad*
West Bengal	*Chakkis*	2-12	—
	Oval ball	1.0	—

The loss in storage when the above packing materials were used was almost nil. The other packing materials available in the market include (a) polyethylene-lined hessian bags and (b) hessian bags with a lining of tar and paper (Khanna and Chacravarti, 1955).

Kaur and Kanwar (1983) suggested storage of *gur* in wooden boxes from April to December. Shinde *et al.* (1983) also confirmed that use of black polyethylene-lined hessian bags were better for *gur* storage. Polyethylene film wrapping prevented running off of the *gur* completely but could not check darkening of its colour. Grass wrapping was found useful for maintaining original colour of *gur.* None of the disinfectants (paraformaldehyde and sulphur dust) could check microbial spoilage of *gur* (Shinde and Marathe, 1981).

Under Gujarat conditions, *gur* blocks with straw and jute cloth before wrapping with black polyethylene sheets is found promising packing material to avoid its deterioration (Patel *et al,* 1991).

Alkathene plus hessian packing for *gur* appeared to be the best packing material since the percentage loss in weight in such packing was only 1.24 per cent as against 46.77 per cent when wrapped in cane trash plus hessian (Mohan Rao *et al.,* 1960).

Packing of *gur* in 150 gauge alkathene film was found to be the most efficient from the point of view of preservation of the physical structure and minimising losses in weight due to the exposure to inclement weather. Wrapping in alkathene film and stitching hessian cloth over it was found to be the most efficient packing material under Andhra Pradesh conditions (Mohan Rao and Lakshmikantham, 1955).

Khanna and Chacravarti (1947; 1949) developed a method for *gur* storage in a blanket of ash. Since there were losses in *gur* weight during storage, they modified this method wherein *gur* blocks stiched in hessian cloth are embedded in ash maintain the product intact in respect of quality and quantity and offers a satisfactory solution of the storage problem. Relative performance of various storage methods (Table 10.5) indicates that certain other moisture proofing materials also produce desired results (Khanna and Chacravarti, 1951). Suitability of this method was also advocated by Prasad and Srivastava, (1954).

Some other materials like alkathene twill bags 110 × 66 cm (with wax bonding between jute fabrics) and alkathene-lined hessian bags (70 × 50 cm) with bitumen bonding were used as packing material for transport purposes but none of the materials were found suitable during transport (Khanna *et al.*, 1957).

Kaur and Kanwar (1986) observed that addition of 100 ppm of sorbic acid effectively checked the deterioration of stored *gur* during monsoon season (July to September). The quality of treated *gur* remained unaffected as it absorbed negligible moisture during storage.

Table 10.5: Relative Performance of Various Storage Methods

Particulars (*gur* stored in in ....)		Fall in Brix.	Fall in Purity	Fall in invert Net rendement	Rise in colour sugar	Fall in trans-mission*
I. Furnace ash	(a) As such	6.36	9.37	16.70	2.72	5.0
	(b) Wrapped in Alkathene bags	0.30	0.22	0.62	0.10	nil
II. Wheat bhusa	(a) As such	0.89	2.48	1.64	0.20	2.0
	(b) Wrapped in Alkathene bags	0.16	0.24	0.48	0.12	—
III. Closed iron drum	(a) As such	0.56	0.74	1.35	0.18	—
	(b) Wrapped in Alkathene bags	0.24	0.24	0.29	0.07	—
IV. Gunny bags	(a) As such	2.66	2.09	3.75	0.34	2.5
	(b) In Alkathene lined bags	0.92	0.43	1.69	0.12	1.5
V. Control (Open trays)	(a) As such	19.98	11.61	26.65	2.94	8.0
	(b) Wrapped in Alkathene bags	0.38	0.40	0.78	0.16	—

* Original transmission being 10 units.
Source: Khanna and Chacravarty, 1951.

The property of a substance to absorb or give out moisture was reported to be governed by its physicochemical composition, particularly its hygroscopic content and moisture (Suryaprakasa Rao and Mohan Rao, 1959). The quantity of moisture of a substance can gain or give up is a function of its geometry of surface area, the difference of the vapour pressure of the substance and its ambient atmosphere, the temperature of atmosphere and time during which it was exposed. Then the permeability of moisture into the packing material also affects the magnitude of the moisture gained or lost by the substance. The empirical form of these relationship may be expressed as :

W = PA t P (at constant temperature),

where

W = moisture gain or loss

P = permeability of water vapour of the packing material, if any

A = Surface area of the substance or of the package used

t = time during which W was recorded

P = difference between vapour pressure of the substance and ambient atmosphere or differences in the vapour pressure inside or outside the packing material when the substance was packed.

REFERENCES

Agrawal, M.P. and A.K. Ghosh (1984a). Designing and evaluation of godown for storing *jaggery* by small farmers. *Indian Sugar,* 34(6) : 1-3.

Agrawal, M.P. and A.K. Ghosh (1984b). External pressure and chemical composition as factors in its keeping quality. *Maharashtra Sugar,* 9(10): 39-42.

Anonymous (Pusa, India) (1953-54). *Annl. Rept.,* Sugarcane Research Station, Pusa, Bihar, India.

Anonymous (1959). Sugar Research in India: 1929-1954—A review III: Sugar Chemistry *Gur* and *Khandsari. Indian J. Sugarcane Res. & Devpt.,* 3 (Pt. II): 109-125.

Baboo, B. and A.K. Ghosh (1985). Effect of gur shapes on moisture absorption. *Co-op.* Sugar 15 (11): 667.

Baboo, B. and A.K. Ghosh (1985). Effect of shapes and artificial drying of *jaggery. Maharashtra Sugar,* 10(5): 47-50.

Baboo, B. and A. Shukla (1987). Improved techniques for domestic storage of *gur* in earthern pots. *Indian J. Sugarcane Technol.,* 4(1) : 27-30.

Baboo, B., P.R. Singh and F.C. Das (1985). Development of *gur* drying-cum-storage bin. *Proc. National Seminar-cum-Group Discussion on Jaggery Manufacture and Storage* held at Indian Institute of Sugarcane Research, Lucknow, India, (Dec. 18-19) : 133-137.

Chacravarti, A.S., S.R. Mukherjee and R.D. Sahi (1969). Storage of *gur* in potato cold storage godown, *Proc. Joint Conv. STAI, DSTA and SISSTA,* 3 Pt. A, A13.1-13.11.

Chouksey, R.G. (1985). Post-harvest technology factor affecting manufacturing and quality of *gur* during storage. *Proc. National Seminar-cum-Group Discussion on Jaggery Manufacture and Storage,* Indian Institute of Sugarcane Research, Lucknow, India, pp. 133-137.

Farooque, M. (1957). Role of microorganisms in deterioration of *gur* and storage. *J. Sugarcane Res. Dev.,* 1(2) : 282.

Gundu Rao, S.N. and N.A. Ramaiah (1961). Deterioration of *gur* during storage (construction of *gur* godown). *Proc. Annl. Conv. S.T.A.I.,* 29 (pt II) : 137-148.

Jadhav, S.K., J.J. Jadhav and D.G. Hapase (1984). The microorganisms associated with *jaggery* (*gur*). *Maharashtra Sugar,* 9(6) : 61.

Kaur, J.R. and R.S. Kanwar (1983). Studies on storage of *gur* in Punjab. *Maharashtra Sugar,* 8(12) : 45-49.

Kaur, J. and R.S. Kanwar (1986). Effect of sorbic acid on the keeping quality of *gur. Indian Sugar Crops J.* 12(2) : 10-11.

Kaur, J.R. and R.S. Kanwar (1990). Keeping quality of *gur* in relation to its inorganic constituents. *Indian Sugar,* 40(4) : 235-237.

Khanna, K.L. (1947) (*c.f.* Anonymous, 1959).

Khanna, K.L. and A.S. Chacravarti (1949). Studies on storage of *gur. Curr. Sci.,* 18 : 127-128.

Khanna, K.L. and A.S. Chacravarti (1951). Improvement of *gur* industry in India-A brief account of researches conducted in Bihar and lines of future work to be followed. *Proc. Bien. Conf. S.R.D.W.I.,* 1 (II) : 74-96.

Khanna, K.L. and A.S. Chacravarti (1955). Scientific Monograph–*Researches on Technical Aspects Relating to Improvement to Gur Industry in Bihar,* Indian Central Sugarcane Committee, New Delhi.

Khanna, K.L., A.S. Chacravarti and M. Farooque (1957). Alkathene-lined jute fabrics as packing material for *gur* in storage and transport. *Indian Sugar,* 6(12) : 731-734.

Lal, U. and R.K. Sharma (1982). Role of shape in its deterioration during storage. *Co-op. Sugar,* 15(1) : 79.

Ligange, A.M., M.R. Hettarchehi and P.M. Jayatissa (1982). Contaminant yeasts of sugarcane products of Sri Lanka. *J. Food Sci. Technol., (Mysore),* 19(6) : 270.

Mohan Rao, N.V. and M. Lakshmikantham (1955). Methods of packaging jaggery for storage–Research note. *Andhra Agric. J.* 2(5) : 256-258.

Mohan Rao, N.V. Mallkarjuna Rao, T. and T. Vankatapathi (1960). Studies in *gur* storage. *Proc. Bien. Conf. S.R.D.W.I.* 4: 427-435.

Panje, R.R. and M.G. Menon (1964). A design of a godown for drying and storing *gur* under North Indian conditions. *Indian J. Sugarcane Res. & Devpt.,* 7(3) : 332-340.

Patel, H.S., N.J. Mehta, H.H. Mistry and M.P. Patel (1991). Effect of different wrapping materials on quality of jaggery during storage, paper presented at the *National Seminar on Increasing Quality and Storability of Jaggery (Gur),* held at the Indian Institute of Sugarcane Research, Lucknow (Nov. 30-Dec. 1), p. 26.

Parthasarathy, S.V. and M. Lakshmikantham (1950). Storage of *gur* in Madras. *J. Agric. Sci.,* 31 (3) : 245-251.

Prasad, S. and D.P. Srivastava (1954). On improved methods of *gur* storage. *Proc. Bien Conf. S.R.D.W.I.,* 2 : 607-612.

Sastry, A.S. and A.K. Ghosh (1965). An assessment of conditions of storage of *gur* in certain trade godowns. *Indian Sugarcane J.*, 8(2): 114-120.

Shakunthala, S.V. (1981). Manual of Sugarcane Production Technology, Department of Agriculture, Tamil Nadu, pp. 142-143.

Shinde, B.N. and S.K. Marathe (1981). Effect of different wrapping materials and disinfectants on keeping quality of jaggery. *Indian Sugar,* 31(9): 609-615.

Shinde, B.N., A.B. Marathe, D.V. Javelekar and S.K. Kadam (1983). Effect of form and colour quality—wrapping material on keeping quality of jaggery during storage. *Indian Sugar,* 32(12): 937-940.

Singh, M., K.M. Bharadwaj and M.L. Agarwal (1978). Storage of *gur*—A review. *Co-op. Sugar,* 9(11) : 559.

Subramaniam, K. and S. Vidyanathan (1964). Preliminary note on large scale *gur* storage under scientific godown conditions. *Indian Sugarcane J.*, 9(1): 40-45.

Suryaprakasa Rao, P.V. and N.V. Mohan Rao (1959). Some fundamental considerations in the storage and transport of jaggery. *Indian J. Sugarcane Res. & Devpt.* III(2): 84-89.

Verma, N.C., K.H. Rao and G.M. Nigam (1958). If you have to hold large *gur* stock. *Indian Fmg.*, 7(8): 23-27.

11

MICROBIAL DETERIORATION OF *GUR*

By and large, *gur* is manufactured in India every year from November to middle of April, and is stored both for marketing and human consumption during the remaining part of the year. *Gur* is mostly spoiled during the monsoon period because invert sugars and mineral salts, being hygroscopic, absorb moisture when ambient humidity is generally high. In other words, moisture is the main culprit in the deterioration of *gur* quality. Due to continuous inversion of sucrose by microbes, more invert sugars are formed which absorb more moisture and this way a noxious circle is formed. The process continues till *gur* becomes soft and loose. At this time, several fungi develop, particularly moulds (aspergilli and penicillia) and members of the family Mucoraceae. The role of moisture in the build up of bacterial population was initially investigated by Farooque (1957).

A review of the available literature shows that deterioration of *gur* can be catalogued into four groups, as follows:

(i) *Physical deterioration* : This is the most common type of deterioration which is due to the darkening of colour. There is also loss of form (shape) through disintegration. Occasionally, the taste is also adversely affected.

(ii) *Chemical deterioration* : This is mostly in the form of inversion and discolouration. In fact, these changes affect the physical characteristics like colour, and shape.

(iii) *Biological deterioration* : It is brought about by insects, particularly ants. Ants feed profusely on *gur*, make tunnels and multiple profusely in these. This type of deterioration is commonly observed in North-eastern hill States.

(iv) *Microbial deterioration* : It has been estimated that about 10 per cent of the *gur* is lost every year during storage due to dryage and deterioration by microbes. Singh *et al.* (1978) critically reviewed the main causes of *gur* deterioration and attributed these to three factors, namely, hygroscopic nature of *gur*, presence of abundant sucrose and high atmospheric humidity. Absorption of moisture is a prerequisite for microbial deterioration as it creates congenial conditions for the development of different types of microorganisms, *viz.*, bacteria, actinomycetes and fungi. The activity of these tiny little organisms results in the evolution of gases, formation of alcohols, organic acids and complex decomposition products besides loss in weight. Because of these changes, *gur* becomes unfit for human consumption.

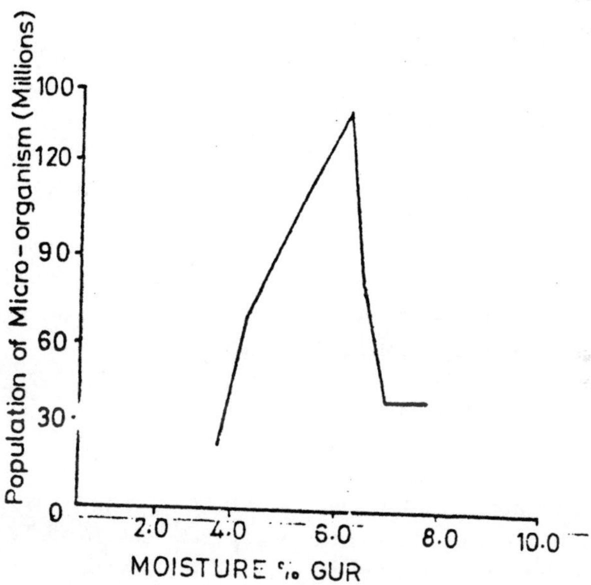

Fig. 11.1: Proliferation of Bacteria in *Gur* at Different Moisture Level (Farooque, 1957)

Deterioration of *Gur* by Bacteria

In humid regions of India, storage of *gur* is really a serious problem and the stockists have to suffer considerable loss of the material. Because *gur* is a hygroscopic substance, its surface quickly absorbs moisture and becomes soft and sticky. At high humidity, the *gur* liquefies completely. The activity of bacteria starts soon after the surface becomes wet. Both spore-forming and non spore-forming bacteria have been isolated from *gur* samples obtained after the monsoon period. Khanna (1952) obtained maximum number of bacteria on the surface (0.6 mm) of *gur* at moisture levels of 10-15 per cent. Farooque (1957) investigated bacterial population in different layers of *gur* blocks. His findings indicated that (i) surface layer of *gur* block (0.6 cm) had the maximum bacterial population, then falling sharply up to a depth of 1.8 cm and beyond this there was no bacterial population (Table 11.1), (ii) at 30°C bacterial population increased with moisture level up to 10 per cent beyond which there was a fall, (iii) critical maximum and minimum levels of moisture in *gur* where bacteria cease to grow at 30°C were 20 per cent and 3 per cent, respectively, and (iv) under storage conditions, the optimum moisture levels for the proliferation of bacteria were between 3 and 6 per cent (Fig. 11.1).

Jadhav *et al.* (1984), for the first time, made concerted attempts to identify the microorganisms, namely, fungi, bacteria and actinomycetes. According to them, among bacteria, population of *Bacillus subtilis* dominated.

Deterioration of *Gur* by Actinomycetes

Actinomycetes, commonly known as 'ray fungi' have certain characters common to both bacteria and fungi, but have distinct characters to delimit them into a separate category. Taxonomically, actinomycetes are closely related to bacteria and have been kept in the order Actinomycetes. On agar plates these can be conveniently distinguished from bacteria. Unlike slimy bacterial colonies, these form tuft and slow growing colonies sticking firmly to agar surface. Under microscope, they display slender unicellular branched mycelia, forming asexual spores for propagation. Actinomycetes proliferate quickly from pH

6.5 to 8.0 and their number declines at pH 5.0. The common genera of actinomycetes are *Streptomyces, Nocardia* and *Micromonospora.* Of these, species of *Streptomyces* always outnumber other forms.

Table 11.1: Bacterial Population in Different Layers of *Gur* after Monsoon

Depth of *gur* layer (mm)	Population (x 10^3)				Mean
	1	2	3	4	
0 - 6.0	373	418	420	428	409.45
6 - 18.5	51	20	21	36	24.60
18.5 - 31.0	Nil	Nil	Nil	Nil	Nil

Source: Farooque, 1957.

Jadhav *et al.* (1984), while investigating the microorganisms associated with *gur,* also reported the presence of actinomycetes as a probable cause for the deterioration and spoilage of *gur* (*jaggery*). According to them, a species of *Streptomyces* was invariably isolated from several *gur* samples analyzed. Agnihorti (1992, unpublished data) also reported association of actinomycetes with *gur* samples. Their population ranged from 4 to 8 × 10^3 per g *gur* sample. Species of *Streptomyces,* on a selective medium, formed off-white colonies while that of *Nocardia* formed brownish colonies.

Deterioration of *Gur* by Fungi

Fungi are cosmopolitan in distribution. These are abundantly present in soil, water and air. Fungal propagules, by and large, are deposited on *gur* samples both during drying, transhipment and storage. All the environmental factors that influence the distribution and proliferation of bacteria and actinomycetes also influence the build up of fungal flora. Fungi, generally, prefer acidic medium. As the pH of the *gur* is generally in the range of 5.5 to 6.5, it becomes an ideal substrate for their growth and reproduction. The following fungi are generally isolated from *gur* samples using dilution plate technique: *Absidia, Alternaria, Aspergilli, Cladosporium, Curvularia, Cylindrospora, Fusarium, Monilia, Mucor, Penicillium, Rhizopus, Spicaria, etc.*

Jadhav *et al.* (1984) also studied the association of fungi with *gur*. They reported that *Aspergilli,* usually outnumbered other fungal species. Among Aspergilli common isclates were: *A. niger, A. humatum., A. flavus, A. claviceps, A. fumigatus, A. ovaluriopsis* and *Eurotium* spp. (perfect stage of *Aspergillus*). Besides aspergilli, they also isolated *Cunninghamella* and *Cylindrocarpon.* Agnihotri (unpublished data) in addition to these fungi also isolated *Aspergillus oryzae, A. tamarii, A. ochraceus, Mucor heimalis, Curvularia* sp., *Penicillium frequentans, Helminthosporium,* sp., *Paeciliomyces* sp., *Rhizopus* sp. and *Trichoderma viride.*

Taxonomic Description of Microbes Isolated from Gur Samples

Alternaria alternata Nees

Colonies fast growing on potato-dextrose-agar at 25 ± 1°C, appressed with scanty aerial mycelium, dark olive-buff to deep olive in shade of dark olive. Reverse dark greenish olive. Mycelium septate, branched 3.4-5.2 µm in diameter. Conidiophores simple or branched, erect, stout, septate, olive brown, straight or slightly curved with geniculations at tip, variable in length, 17.0-85.0 µm × 3.4-5.1 µm. Muriform conidia produced in chain, olive brown to dark brown, cylindrical, obclavate to rather short beaked, with 2-7 transverse septa and 3-5 longitudinal septa generally produced in third or fourth cell, 13.6-57.8 µm × 6.8-17.0 µm, average 27.2-34.0 µm × 10.2-13.6 µm with beak.

Aspergillus fumigatus Fresenius

Colonies on Czapek's agar spreading broadly, velvety, reverse colourless. Conidial heads long, compact, when young turning into dirty green in aged ones up to 400 µm but smaller heads also present. Conidiophores crowded, smooth, short, up to 300 µm in length by 4.2-8 µm in diameter. Vesicle flask shaped, usually fertile over the upper half, 20-30 µm in diameter. Sterigmata in one series, closely packed, ranging from 6-8.7 µm by 1.5-2.8 µm in diameter. Conidia globose, echinulate, 3-3.5 µm in diameter.

Aspergillus flavus Link

Colonies on Czapek's agar floccose, spreading rapidly and attaining a diameter of 4-5 cm in 8-10 days at room temperature (28 ± 2°C), surface velvety, zonate, light green at first then changing to parrot-green and finally to darkish green with age, reverse orange yellow. Conidial heads green, globose to loosely radiate, regularly splitting into 2 to 4 columns, 300-420 μm in diameter, arising from submerged hyphae, pitted, colourless. Vesicle globose to subglobose, 12-35 or 4 μm in diameter. Sterigmata usually in two series but may be in single series on smaller heads. Primary sterigmata 7-10 μm by 2.5-3.0 μm, secondary 7-10 μm by 2.5-3.0 μm in diameter, rough. Conidia almost globose, echinulate, 3.0-4.5 μm in diameter. Sclerotia at first white, then brownish-black with white tips. Perithecia not observed.

Aspergillus nidulans (Eidam) Winter

Colonies on Czapek's agar growing moderately at room temperature attaining a diameter of 3-4 cm in 10-12 days, velvety, dark cress green with age. Heads short, globose ranging from 60-70 μm by 30-35 μm. Conidiophores arising directly from submerged hyphae, smooth-walled with brownish tinge, 51-128 μm by 2.5-3.5 μm, average 70 μm by 3 μm. Vesicle hemispherical, 8-15 μm in diameter, fertile over the upper half. Sterigmata in two series, primary 4.5-5.4 μm by 2.8 μm, secondary 6-6.5 μm by 2-2.2 μm. Conidia globose, rugulose, green in mass, 2.8-3.2 μm in diameter. Cleistothecia large, abundant, dark pink, ranging from 90-165 μm in diameter. Cleistothecial wall many layered when young but at maturity single layered in thickness. Hulle cells globose, thick-walled, usually 16-22 μm in diameter. Asci globose, 7-10 μm in diameter. Ascospores 8 in each ascus, purple red, lenticular, smooth walled with two equatorial crests, ranging 3.8-4.5 μm in length by 3.5-4.0 μm in breadth. Equatorial ridges two in number, 0.5-0.7 μm in diameter.

Aspergillus niger van Tieghem (Sclerotial strain)

Colonies on Czapek's agar rapidly growing with abundant

submerged mycelium at 25°C attaining a diameter of 4-5 cm in 10-12 days. Reverse dirty-white to slightly yellowish. Conidial heads blackish-brown to purple-brown, globose to radiate, up to 250 μm in diameter. Conidiophores mostly arising from the substratum, septate or with occasional septa, 400-600 μm in length by 7-12 μm in diameter, walls yellowish. Vesicle globose or subglobose, thick-walled, fertile over the entire surface, 28-40 μm in diameter, yellowish brown in colour. Sterigmata in one series in smaller heads but mostly in two series, more or less brown, closely packed, primary 18-30 μm by 6-7.5 μm; secondary 6.0-9.2 μm by 1.5-2.5 μm. Conidia echinulate, brownish-black, globose, 2.5-3.6 μm in diameter. Sclerotia abundantly produced, white, later becoming light pink, 400-600 μm in diameter.

Aspergillus oryzae (Ahlburg) Cohn

Colonies on Czapek's agar rapidly growing at 25 ± 1°C attaining a diameter of 4-6 cm in a week. Hyphae mostly submerged and forming a white to grey mycelial layer in the form of a felty mass. Conidial heads at first white, then yellowish-green and finally dark green, ranging from 500-600 μm in diameter, on maturity splitting into columns. Conidiophores 1000-1500 μm in length by 13-15 μm in diameter, walls thin and pitted. Vesicles globose to subglobose, 30-58 μm in diameter with wall 1-1.5 μm in diameter. Sterigmata commonly in one series ranging from 8-15 μm by 3-5 μm in diameter, or in two series with primary sterigmata up to 12 by 5 μm. Secondary sterigmata 10-12 μm by 3.5 μm. Conidia more or less pyriform, 5-6 μm by 3.5 μm. Sclerotia first white turning black in old cultures.

Aspergillus rugulosus Thom and Raper

Colonies on Czapek's agar growing restrictedly and attaining a diameter of 2-3 cm in 8-10 days. Conidial heads green short, columnar, 70-100 μm by 30-40 μm in diameter. Conidiophores sinuous, smooth-walled with brownish tinge 50-80 μm long, varying up to 5 μm in width, gradually enlarging towards the vesicle, 8-10 μm in diameter. Sterigmata in two series, primary sterigmata 7-8 μm by 3-3.6 μm, secondary sterigmata 6-7 μm by 2.5-3.4 µm in diameter. Cleistothecia in abundance, brown in colour becoming purple gray to purple-brown with age, globose, 22-350 µm in diameter, full of asci. Asci 10-11 µm in long axis. Ascospores purple-red, wall rugulose, lenticular, 4-4.4 µm by 3.6-3.8 mm with two pleated equatorial crests. Conidia globose, rugulose, 3-3.5 µm in diameter. Hulle cells abundant with thick walls, 15-22 µm in diameter, surrounding the cleistothecia. Cleistothecia scattered, wall brittle, single layered. Asci many in cleistothecium, their wall breaking quickly to leave the cavity filled with ascospores. Ascospores purple-red, star shaped, with spore bodies lenticular 3.5-4 µm by 3 µm, with two equatorial ridges which are up to 3.5 µm in width.

Asperigillus tamarii Kita

Colonies on Czapek's agar spreading rapidly attaining a diameter of 5-6 cm in 10-12 days, vegetative mycelium almost submerged. Conidial heads at first white then passing through orange-yellow shades to brown in old cultures. Conidial heads abundant, brown, globose, up to 300 µm in diameter with 3-5 columns of conidia. Conidiophores arising directly from the substratum, colourless, septate, conspicuously rough, ranging from 600-800 µm in length by 9-12 µm in diameter, gradually broadening towards the vesicle. Vesicle globose to subglobose, 25-35 µm in diameter, fertile over almost the entire surface. Sterigmata in one series in small heads, in two series in large heads; primary sterigmata commonly 7-10 µm by 3-4 µm, secondary 8-9 µm by 2.5-3 µm in diameter. Conidia globose to subglobose, conspicuously roughened, brown in mass, 4-6 µm in diameter. Sclerotia not seen.

Aspergillus terreus Thom

Colonies on Czapek's agar growing moderately at room temperature (20 ± 1°C), reaching a diameter or 3-4 cm in 10-12 days, buff in colour, velvety, reverse in brown shades. Exudates in shining droplets. Conidial heads globose to columnar up to 24 µm in length. Conidiophores, colourless, more or less flexous smooth, commonly ranging from 100-250 µm by 5-6 µm in diameter. Vesicle dome shaped, 10-16.2 µm in diameter, sterigmata in two series primary 5-6.8 µm, secondary 5.5-8.0 µm by 2 µm in diameter. Conidia globose commonly 1.8-2.8 µm in diameter.

Cunninghamella echinulata Thaxter

Colonies on potato-dextrose-agar fast growing at 25 ± 1°C, turf white changing to yellowish with age, smooth, floccose. Mycelium non-septate, interwoven, branched, hyaline, 3.0-6.0 µm wide. Conidiophores erect, more or less irregularly and indefinitely branched, non-septate variable in length. Terminal vesicle variable in length, ovate, nearly spherical to ovoid, 34.0-74.8 µm in diameter with an average 40.8-51.0 µm; lateral branches present in two to three whorls; lateral vesicles more or less similar but smaller in diameter 17.0-34.0 µm in diameter. Conidia oval to elliptical, colourless finely echinulate, 7.5-19.5 µm × 7.5-15.0 µm.

Cladosporium cladosporioides (Fres.) de Vries

Colonies on Czapek's agar growing slowly, attaining a diameter of about 2.5 cm in 7 days at 25 ± 1°C, olive-green, velvety to slightly flocose, central portion slightly raised. Reverse greenish black. Conidiophores macronematous as well as micronematous, 35-310 µm in length, 3.3-4.4 µm in diameter, branched smooth. Ramo-conidia 0-1 septate, 10-30 µm × 3.3-4.4 µm, smooth or slightly verrucose conidia formed in long branched chains, non-septate, ellipsoidal to lemoniform, 3.3-9.9 µm × 2.8-4.0 µm, olive-green, smooth.

Curvularia lunata (Walker) Boedijn

Colonies on potato-dextrose-agar fast growing, smooth, first white becoming dark greyish olive to black due to heavy sporulation. Reverse olivaceous-black. Mycelium light brown, branched 3.4-6.8 µm wide. Conidiophores septate, simple geniculate at the tip, brown 17-112.2 µm × 3.4-5.1 µm. Conidia mostly three septate but sometimes 1-5 septate, the third cell from base larger in size and darker in colour than other cells, pale-brown, clavate or pear shaped, nearly straight to more or less curved, constrictions in septa not prominent, 15.7-29.5 μm × 6.9-12.8 μm. Chlamydospores present, more or less spherical, dark-brown 10.2-17.0 μm, average 13.6 μm in diameter.

Fusarium moniliforme Sheldon

Colonies white on potato-dextrose-agar, compact, appearing powdery with production of abundant microconidia. Microconidia in long chains, dominating over macroconidia, small, 0-1 septate, hyaline, 3.0–16.5 μm × 1.5–4.5 μm. Macroconidia thin walled, scattered, formed mostly in sporodochia or pinnotes, sabulate to cylindrical, slightly curved, narrowing at both ends, brownish white in mass, 1-5 septate but mostly 3-septate, 15-60 μm × 4.5-6 μm.

Helminthosporium tetramera Mckinney

Colonies on potato-dextrose-agar compact, growth felty, quaker darb, deep quaker darb to dark greenish-olive. Mycelium septate, branched, dark olivaceous, 3.4-6.8 μm wide. Conidiophores abundant, dark olivaceous to almost brown, septate at the base, geniculate irregularly at tips, 40.8-272 μm × 6.8-8.5 μm. Conidia strictly four celled, olivaceous-brown to dark brown, symetrical, almost straight towards the rounded ends, borne in cluster, 13.6-37.4 μm × 10.2-13.6 mm, average 23.8-27.2 μm × 10.2 μm.

Mucor hiemalis Wehmer

Colonies on potato-dextrose-agar fat growing, white, cottony, changing to greyish-yellow in old age. Reverse colour buff yellow. Sporangiophores erect, unbranched, sporangia spherical to globose, bright yellow with age, 23.8-81.6 μm, average 51.0 μm in diameter. Columella spherical to oval, free, colourless, spherical 13.6-34.0 μm, average 27.2 μm in diameter oval 20.4-34.0 μm × 17.0-30.6 μm, average 27.2 × 23.8 μm. Spores unequal, mostly elongate, ellipsoid or kidney shaped, 3.0-7.5 μm × 1.5-6.0 μm, hyaline, smooth. Chlamydospores 10.2-26.2 μm. Zygospores absent.

Paecillomyces varioti Bainier

Colonies on Czapek's agar fast growing, spreading, floccose, olive buff to brownish-olive, powdery in appearance, reverse colourless. Conidiophores arise from submerged mycelium, erect, freely and irregularly branched, 68.0-244.8 μm × 6.8-13.6 μm, variable in size, each branchlet bears divergent metulae. Metulae of variable length varied from 23.8-51.0 μm × 5.1-6.8 μm. Phialides divergent, 9-24 μm × 2.25-4.5 μm. Conidial chain vary, long, conidia smooth, pale olive-buff to olive-buff, oval to elliptical, mostly elliptical 4.5 -6.0 μm × 3.0-4.5 μm.

Penicillium frequentans Westling

Colonies on Czapek's agar fast growing, attaining a diameter of 2-3 cm in 10 days at 25 ± 1°C, velvety, marginal areas slightly floccose, radiately wrinkled, central portion sulcate. Mycelium closely woven, hyphae either above or below the substratum from which abundant crowded conidiophores arise, heavily sporing throughout in artemisia green shape, changing to state olive in age, no exudate, odour faint (mouldy), reverse colourless at first to yellowish orange, becoming more brown in old cultures. Conidiophores short, arising directly from the substratum, ranging from 75-140 μm in length by 3.4-3.8 μm in diameter, with walls smooth to finely roughened, apices enlarged up to 5.5 μm in diameter; conidial structures forming a crowded, fairly continuous layer over the entire colony surface, with penicillin strictly monoverticillate. Sterigmata 5-10 in each verticili, measuring 8.4-12.6 μm by 2.0-2.8 μm. Conidia globose to subglobose, delicately roughened, ranging from 2.5-3.8 μm in diameter, forming compact conidial columns up to 165 μm in length.

Rhizopus oryzae Went and Gerlings

Colonies on potato-dextrose-agar fast growing at room temperature (25 ± 1°C), turf thick, fetty white, changing to greyish white later dark brown. Mycelium hyaline becoming brown with age, septate, branched with well developed rhizoids and stolons. Sporangiophores arise opposite to rhizoids in groups of 2-5, rarely single, hyaline to brown, 347.3-1963 µm × 7.55-22.65 µm. Sporangia terminal, hemispherical, brown columellate, many spores, 45.3-151.0 µm in diameter. Columella well developed, yellowish-brown, spherical to hemispherical, 22.65-30.2 µm in diameter. Sporangiophores oval to slightly elliptical, pale olive-buff to olive-buff, 4.5-7.5 µm × 3.0-6.0 µm. Chlamydospores and zygospores not observed.

Trichoderma lignorum (viridi) (Tode.) Harz.

Colonies on potato-dextrose-agar, fast growing at room temperature (25 ± 1°C), thick and appressed in the centre with smooth margin, pale blue green in centre and deep malachite green at the margin. Reverse colourless in shades of yellow to cream buff. Mycelium thin, bright, light fluorite green, septate 3.0-7.5 µm in diameter. Conidiophores septate, 24.0-67.5 µm × 3.0-3.75 µm, dichotomously or trichotomously branched, occasionally forming whorls, branches swollen at the base and pointed towards the apex, measuring from 7.5-15.0 µm × 1.5-3.0 µm with an average of 2.25 µm in diameter. Conidial heads fluorite green, 9-12 µm in diameter. Conidia produced in clusters on the tips of a branched conidiophore, globose to ovate 2.25-3.75 µm, average 3.0-2.75 µm, smooth, light fluorite green in colour.

Preserving *Gur* Against Deterioration

Deterioration in *gur* quality *per se* occurs at the following stages:

(a) harvesting unripe and stand-over canes
(b) time lag between harvesting and crushing
(c) time lag between crushing and boiling of juice for making *gur,* and
(d) Storage conditions of *gur.*

Magnitude of deterioration further increases if the cane is from problem soils (saline or waterlogged), has received excessive application of nitrogenous fertilisers, suffered frost damage or it is affected by pests and diseases. It may be mentioned here that use of a deteriorating variety further complicates this situation. Sugarcane varieties are known to differ in almost all the above attributes and stages that influence deterioration. In open-pan system of boiling, juice is generally left in *mattas* and pans for 4 to 5 hr usually at 25-30°C and as a result of inversion reducing sugars are formed. Around 2.37 per cent of the available sugars are lost in juice when stored for 5 hr at 28°C. This loss may be still more if the juice is obtained from stale canes (containing more of reducing sugars) and kept at higher temperatures (Hadi, 1929).

Application of phosphorus (as fertiliser) has been acclaimed to reduce deterioration in the keeping quality of cane after attaining peak maturity (in terms of brix) (Rege and Sannabhadti, 1943).

For overcoming post-harvest losses, waxing paraffin of the cut ends was effective in preventing deterioration in juice quality at least for 24 hr and thereafter deterioration appeared to be rather quick in those canes where cut ends were not waxed (Barnes, 1918). Loss of available sugars in the juice of sugarcane varieties S.48 and Co 221 was 0.32 and 0.35 per cent, respectively for the canes stored in shade and 0.6 per cent for those stored in the sun (Hadi, 1929). Spraying water in the form of mist during loading of canes in wagons/carts/trucks with 10 litres of water/t cane ensured 0.5 to 0.7 unit increase in recovery (Khanna *et al*., 1953). Storing canes in shade and spraying water in the morning and evening can also reduce deterioration of juice to some extent.

Studies conducted at the Sugarcane Breeding Institute, Coimbatore, indicated that dipping cut end of canes immediately after harvest in 200 ppm Bactrinol-100 solution for 5 min minimised staling losses up to 120 hr (Anon. (Coimbatore, India), 1987). Dipping cut ends of canes in/or spraying B_D Mill Sanitizer (1000 ppm) prevented deterioration of canes effectively up to 24 hr and thereafter slowed the pace of deterioration (Srivastava *et al.*, 1989). Mixing Polycide (2500 ppm) and Bactrinol-100 (500 ppm) in juice prevented its deterioration up to 24 hr. *Gur* prepared from such a juice could be made even after 48 hr, with residues below the detection range (Anon. (Coimbatore, India), 1987). Quality of this *gur*, however, was poor as Bactrinol-100 did not check the inversion of sucrose which resulted in increase in reducing sugars and the *gur* became hygroscopic (Chiranjivi Rao and Thangavelu, 1987).

Dipping harvested cane in Actin ID solution (0.15%) or Leukokil (1.0%) was effective in minimizing sugar losses. Pre-harvest foliar application of cobalt chloride, barium chloride or zinc sulphate (100 ppm) considerably checked the inversion of sugars in harvested canes (Solomon, 1990). Mixing IFOPOL (10-100 ppm) in juice fully inhibited the formation of oligosaccharides (which diminish processing efficiency and raw sugar quality) for more than three days (Susana *et al.*, 1991a). Rivero and Ramos (1991) observed that using IFOPOL @ 1 l -1000 t^{-1} juice reduced the content of both polysaccharides and oligosaccharides by 3.5 times. Formol has also been recommended as antimicrobial disinfectant for maintaining hygienic conditions in the sugar plants in Cuba (Susana *et al.*, 1991b).

For preservation microbial deterioration of *gur* during storage, fumigation by sulphur containing compounds (Suryaprakasha Rao and Mohan Rao, 1959) or smoking by burning rice husk has been found useful (Sastry and Ghosh, 1965).

REFERENCES

Anonymous (1959). Sugarcane research in India : 1929-1954—A review. III. Sugarcane Chemistry, *Gur* and *Khandsari*. *Indian J. Sugarcane Res. and Devpt.*, 3(pt.II) : 109-125.

Anonymous (Coimbatore, India) (1987). Research Achievements; 1912-1987, Platinum Jubilee Profile (Ed. David, H.), Sugarcane Breeding Institute, Coimbatore, pp. 121.

Barnes, J.H. (1918). Sugar and the Sugarcane in the Gurdaspur District, Agricultural Research Institute, Pusa (Bihar), *Bulletin No. 69*, pp. 100, Supdt. Govt. Printing, India, Calcutta.

Chiranjivi Rao, K. and S. Thangavelu (1987). Bactronol-100- A potential and promising biocide for sugar industry. *Proc. Annl. Conv. S.T.A.I.*, 60 : 627-634.

Farooque, M. (1957). Role of microorganisms in deterioration of *gur* in storage. *Indian J. Sugarcane Res. and Devpt.*, 1(2) : 69-71.

Hadi, S.M. (1929). The Indian Sugar Industry, Govt. of Bhopal, C.I., printed by F.W. Petch Thacker and Co. Ltd., Bombay, published by Khan Bahadur S.M. Hadi, Director Agriculture, Bhopal State, pp. 289 + vi.

Jadhav, S.K., J.J. Jadhav and D.G. Hapase (1984). The microorganisms associated with jaggery (*gur*). *Maharashtra Sugar*, 9-(6) : 61-66.

Khanna, K.L. (1952). *Annl. Rept.* Sugarcane Res. Stn., Pusa Bihar, India.

Khanna, K.L. *et al.* (1953) . (*c.f.* Anon., 1959).

Rege, R. D. and S. Sannabhadit (1943). Problem of sugarcane physiology in the Deccan Canal Tract IV Mineral nutrition : (A) Phosphorus. *Indian J. Agril. Sci.,* 13 (1) : 87-111.

Rivero, A. and E. Ramos (1991). Disinfection facility in sugarcane cleaning station. *Cuba Azucar* (October-December) : 7-9.

Sastry, A.S. and A.K. Ghosh (1965). Effects of paddy husk smoke on the preservation of *gur*. *Indian Sugar J.,* 9(4) : 234-241.

Singh, M., K.M. Bharadwaj and M.L. Agrawal (1978). Storage of *gur*—A review. *Co-op. Sugar,* 9 (11) : 559-567.

Solomon, S., K.K. Srivastava, S. Bhatnagar and V.K. Madan 9(1990). Post-harvest changes in invertase activity and juice quality in sugarcane. *Indian Sugar,* 39 (12) : 895-899.

Srivastava, M.K., S.P. Shukla and A.P. Dixit (1989). Measures to check sugarcane post-harvest deterioration. *Proc. Annl. Conv. S.T.A.I.,* 52 : Ag. 78-Ag. 90.

Suryaprakasha Rao, P.V. and N.V. Mohan Rao (1959). Some fundamental considerations in the storage and transport of jaggery, *Indian J. Sugarcane Res. and Devpt.,* 3 (1-4) : 84.

Susana, R.B., L.R.S. Eduardo and M.T.P. Barbara (1991a). Inhibition of oligo- and polysaccharide formation during sugarcane deterioration. *International Sugar J.,* 93: 180-183.

Susana, A.D., O.R. Cruz, X.M. Leal and M.H. Leal (1991b). Antimicrobial action of Formol on sugarcane juices. *Cuba Azucar* (October-December).: 32-37.

12

GUR FROM PEST AND DISEASE AFFECTED CANE

Insect pests and diseases are important biotic stresses influencing cane and sugar productivity. These biotic stresses may affect productivity of *gur* and its quality as given below:

(i) Pests and diseases reduce crop productivity per unit area, average cane weight and quality, resulting into commensurate decrease in *gur* production.

(ii) The extraction of juice from canes affected with pests and diseases is reduced.

(iii) Pests and diseases cause deterioration of cane and permit build up of certain undesirable substances (like nitrogen, dextrans, reducing sugars, *etc.*) which adversely affect quality of *gur*.

Effect of Important Pests on Quantity and Quality of *Gur*

Sugarcane crop is attacked by at least 125 species of different insect pests. Out of this number, several lepidopterous pests like, stalk borer, shoot borer and Gurdaspur borer and sucking pests like *Pyrilla*, black bug and scale insects cause enormous loss to sugarcane crop. The impact of these pests is several times more in the ratoon crop which is generally used for making *gur* early in the season. Although considerable information is available on the impact of these pests on sugar recovery (Table 12.1), limited studies have been made on the production of *gur* from pests affected sugarcane. This type of information is essential because *gur* plays a pivotal role in rural economy. Information on the effect of various insect pests on cane yield and recovery is presented below (Tables 12.1 and 12.2). The losses increase with increase in the level of incidence of a particular pest (Table 12.2).

Table 12.1: Effect of Pest Infestation on Yield and Quality of Sugarcane

Insect-pest	Loss in				Reference
	Cane yield	Av. cane wt.	Sucrose	Recovery	
Chilo simplex	—	—	1.65-3.24%	—	Hadi (1929)
Chilo auricillius (Stalk borer)	—	—	1.7-3.6%	2 units	Avasthy (1973)
C. tumidicostalis (Plassey borer)	35-100 t ha^{-1}	—	—	3 - 26%	Avasthy (1973)
C. indicus (Inter-node borer)	3 -4 t ha-1	—	—	1.12%	David and Kalra (1965)
C. infuscatellus (Shoot-borer) per unit intensity	2.5 t ha^{-1}	—	1 - 4 units	2 units	Avasthy (1973)
Emmalocera depressella (Root borer)	10%	—	—	0.3 units	Gupta and Avasthy (1952)
Melanaspis glomerata (Scale insect)	—	25.0%	6.5%	—	Moholkar *et al. (1973)*
	—	43.0%	8.03%	—	Khanna (1957)
Aleuro barodensis Mark (Termite)	33.0%	—	—	1-4 units	Gupta and Avasthy (1952)
Pyrilla perpusilla (Pyrilla)	—	—	30 - 40%	—	Kalra (1983)
Odontotermes obsesus (White fly)	—	—	1.98 - 2.54%		Avasthy (1973)

Effect of Stalk Borer

In north India, *C. auricillius* (stalk borer) is an important pest of sugarcane. The pest attacks the crop in U.P. and Bihar from July to December and its incidence has been correlated with the growth and thickness of the cane. It inflicts heavy losses both in yield of cane and sugar. The stalk borer is comparatively more harmful than other borers as it appreciably reduces the sucrose content in juice. Nigam (1960) reported that due to this borer, the cultivators suffered a loss of 15 per cent in yield and

2 units in sucrose per cent juice. Another estimate by Gupta and Singh (1951) puts the mill loss at 17 per cent in the weight of the damaged cane and 1.69 to 3.66 per cent in sucrose.

Table 12.2: Loss due to Insects Pests at Various Levels of Incidence in Uttar Pradesh (Subtropical India)

Pest	Level of incidence (% or otherwise)	Loss in	
		Cane yield (%)	Recovery(units)
Termite	1.4	2.88	—
	7.4	14.85	—
	11.1	26.06	—
	21.8	28.35	—
Root borer#		2.5	—
Top borer	Early attack	74.3	5.66
	Attack in the mid-season	11.3	0.60
Stalk borer	1 - 10	4.5	0.18
	11 - 25	6.5	0.79
	26 - 50	15.5	1.73
	51 - 75	27.7	4.20
White fly	Light		1.21
	Moderate		2.71
	Heavy		2.80
Scale insect	Moderate	8.33	0.97
	Heavy	11.66	3.08
White grub and beetle	Moderate	—	1.7
	Heavy	44.3	3.65
Pyrilla	Severe	25.0	2.3

during monsoon and post-monsoon period.
Source: Based on the information given in Anon., (Shahjahanpur, India), 1983.

Gupta and Kulshreshtha (1960) have reported an average loss of 139.56 quintal per ha as a result of borer damage. They further reported that due to borer attack the average decrease in quantity of *gur* and *khandsari* was 2.8 per cent and 20.22 per cent, respectively, amounting to the reduction in total production by 14.67 q/ha in *gur* and 4.84, 1.35 and 0.43 q/ha in first, second and third *khandsari* sugar, respectively.

A review of the work done at Shahjahanpur (Anon., (Shahjahanpur, India), 1983) indicated that per cent loss in *gur* at 5-17.6, 21.4-40 and 45.8-100 per cent infestation on internode basis was recorded to be 0.29, 4.09 and 19.88 per cent, respectively. Further, there was some deterioration in *gur* quality also.

The data presented in Table 12.3 amply show that stalk borer infestation appreciably reduces the production of *gur* and different grades of *khandsari* sugar when compared with the normal recovery. The losses were variety-oriented as Co S 416 suffered the least in *gur* and *khandsari* production in spite of heavy borer infestation.

Table 12.3: Influence of Stalk Borer Incidence on Recovery of *Gur* and *Khandsari* Sugar

Variety	Borer incidence	Recovery of *gur* (%)	*Gur* yield (qt/ha)	*Recovery of khandsari* sugar (per cent)		
				1st	2nd	3rd
Co 312	99.8	8.35	4.97	3.22	0.88	0.29
Co 453	89.2	6.43	5.89	2.83	0.77	0.28
Co 527	94.8	7.77	7.69	2.78	0.76	0.25
Co 617	96.0	8.77	7.34	2.78	0.76	0.25
Co 416	85.2	9.17	4.46	9.95	1.35	0.45
CoK 41	98.2	4.02	4.84	1.74	0.47	0.17
Mean	93.8	7.41	5.80	3.05	0.83	0.28
Normal recovery	—	10.00	—	3.67	1.00	0.33

Source: Gupta and Kulshreshtha, 1960.

Singh and Pandey (1967) made critical studies on borer incidence *vis-a-vis gur* production (Table 12.4). They reported that practically there was no loss in cane juice.

When the borer incidence was up to 29.2 per cent on the internode basis, but with further increase in incidence the juice loss also increased. At 69.2 per cent borer incidence, the juice extraction was only 46.1 per cent against 52.7 per cent in the healthy control. The weight of *gur* produced also decreased with increase in incidence of stalk borer attack. The loss in *gur* being 0.29, 4.09 and 19.8 per cent, respectively in the three categories mentioned above.

Singh and Pandey (1967) also analysed the *gur* for various physical and chemical characteristics. They reported that there was little difference in different qualities of *gur* prepared from the juice extracted out of healthy and the cane showing 12.1 per cent average incidence of the stalk borer on internode basis.

The quality of the *gur* determined was significantly affected at 29.2 and 69.2 per cent borer incidence.

Table 12.4: Influence of Borer Incidence on Juice and *Gur* Production

Character	Per cent incidence (internode basis)			
	5 - 17.6	21.4 - 40.0	45.8 - 100.0	Healthy
Average incidence (internode basis)	12.1	29.2	69.2	—
Cane wt. (kg)	25.4	25.6	2.8	25.4
Juice wt. (kg)	13.4	13.5	11.9	13.4
Extraction (%)	52.7	52.7	46.1	52.7
Juice used for *gur* making (kg)	10.0	10.0	10.0	10.0
Wt. of *gur* (kg)	1.70	1.64	1.37	1.71
Gur per cent juice	17.0	16.4	13.7	17.1
Gur per cent cane	8.9	8.6	6.3	9.0
Loss of *gur* over healthy	0.29	4.09	19.88	—

Source: Singh and Pandey, 1967.

Avasthy (1973) also reported that quality *gur* cannot be prepared from juice of shoot borer, another lepidopterous pest, infested crop.

Effect of Scale Insect

Sugarcane crop is damaged by 35 species of scale insects. Of the different species, *Melanapsis glomerata,* the native scale insect, is economically the most important. The pest is invading new areas through uninterrupted movement of seed material from one location to the another. This is the reason that the pest occurrence is now being reported from areas where it was hitherto unknown. In recent years, scale insect epidemics have occurred in several sugarcane growing States of the country posing serious threat to sugarcane cultivation. Scale insect not only reduces the germination and cane yield (Khanna, 1957) but also reduces percentage of juice extraction, its quality and *gur* production. The quality of *gur* depends on the population of scale insect in the infected cane, stage and type of crop, sugarcane variety, *etc.*

According to Prabhakara Rao *et al.* (1976), the *gur* prepared from the scale insect infested crop does not set properly. Besides reduction in *gur* production, the *gur* produced is of dark colour and is high in reducing sugars (> 10%); this reduces the market value of the produce (Moholkar and Ranadive, 1973). Moholkar *et al.* (1973) made detailed studies on the effect of scale insect infestation *vis-a-vis gur* production. They reported 27.0 per cent decrease in *gur* recovery which was ascribed to decrease in sucrose (6-8%) and increase in N content (29.0%) in the juice. The *gur* so prepared contained less sucrose (81.6% as against 83.6% in normal), and relatively more of reducing sugars (11.6% as against 10.7% in normal) and colour intensity (79.0 units as against 69.0 units in normal canes).

It has been reported that when canes are given hot water treatment (HWT) for 3-5 min before crushing the colour of the *gur* appreciably improved (Anon., 1950). Smith (1975) reported the blanching of cane (treatment of canes at 80-90°C for 10-15 min) reduces the colour of the cane juice by 47 per cent. Besides, this benefit, the other potential benefits are less formation of invert sugars because of heat deactivation of invertase, less foaming, fewer settleable solids in clarification and improvement in filterability. Prasada Rao *et al.* (1984) also reported that *gur* made from mature canes blanched with hot water was found to be better in colour, hardness and had lesser moisture content. Keeping these benefits in view Ramakrishna Rao *et al.* (1986; 1991) studied the production of *gur* from hot water treated scale insect infested canes. These workers reported that *gur* produced from scale insect infested cane of Co A 8201 (early ripening variety) subjected to blanching at 80-90°C for 20 min or infested cane was washed with cold water gave significantly harder *gur* as compared to that obtained from scale infested cane (Table 12.5).

The improvement in quality of *gur* could be due to deactivation of certain enzymes as there was no effect on sucrose, reducing sugars and non-sugars present in *jaggery* (Ramakrishna Rao *et al.*, 1991).

At Melalathur, in Tamil Nadu, soft and high sugar varieties are susceptible to another species of scale insect *Marsipococcus* spp. especially under waterlogged conditions from the month of July to December. Chitinous covering of these insects are also crushed along with the cane giving very turbid juice. Even if it is filtered 2 to 3 times, the *gur* prepared is dark brown in colour and deteriorated quickly if stored. A loss of 1.6 per cent in jaggery recovery was observed in the *gur* prepared from infested cane (Shakunthala, 1991).

Table 12.5: Effect of Hot Water Treatment (HWT) and Scrubbing with Cold Water of Scale Insect Infested Cane on Composition of *Gur*

Particular	Scale infested cane	Blanching of cane				Scale infested cane washed with cold water
		50 - 60°C		80 - 90°C		
		10 min	20 min	10 min	20 min	
Hardness (in cm under 5 kg pressure)	1.38	1.32	1.28	1.09	0.75	0.86
Colour reading (N/2 solution)	396	398	379	286	215	239
Moisture (per cent)	4.85	5.09	4.78	4.50	3.68	3.98
Sucrose (per cent)	76.68	76.59	76.75	76.92	77.25	77.05
Reducing sugars (per cent)	7.25	7.05	7.10	6.95	6.90	6.85
pH (N/2 solution)	6.15	6.15	6.20	6.15	6.15	6.15
Total non-sugars (per cent)	12.96	13.05	12.94	13.05	12.75	12.85

Source: Ramakrishna Rao *et al.*, 1986.

Effect of Pyrilla *Infestation*

Among the sucking pests of sugarcane, *Pyrilla perpusilla,* a fulgorid leaf hopper, is the most destructive. The pest occurs throughout the length and breadth of the country, but its menace is more in north India as compared to south India. Both adults and nymphs of the pest suck cell sap from the underside of the leaves near the mid rib. Due to excessive feeding the leaves turn yellow and while sucking the sap the pest secretes a 'honey dew' like substance on which *Capnodium* spp. grow luxuriantly. The black mycelial growth of the fungus interferes with the photosynthesis of the leaf impairing carbohydrate metabolism of the host plant. A severe infestation of *Pyrilla* causes loss of 30-40 per cent in sucrose and 20-25 per cent in total salts; this causes deterioration in juice quality.

Duhra and Agarwal (1987) made in depth study on the level of infestation of *Pyrilla* in relation to juice quality of Co J 64, a high sugar early ripening variety (Table 12.6).

Table 12.6: Effect of *Pyrilla* on Juice Quality of Sugarcane (Var. Co J 64)

Particular	Sampling date								Per cent decrease (–)/ increase (+) over healthy
	Healthy cane				Infested cane				
	Oct.	Nov.	Dec.	Mean	Oct.	Nov.	Dec.	Mean	
Brix (%)	18.18	19.36	19.12	19.12	13.78	15.32	18.14	15.75	(–) 17.6
Sucrose (%)	14.84	16.58	17.28	16.23	9.58	11.32	15.09	12.06	(–) 25.5
Non-sugars (%)	3.34	2.78	2.54	2.89	4.20	4.00	2.85	3.68	(+) 27.3
Purity (%)	81.60	85.68	87.18	84.82	69.61	73.73	84.29	75.87	(–) 10.5
CCS (%)	8.91	10.21	10.73	9.95	5.21	6.41	9.34	6.90	(–) 29.7

Source: Duhra and Agarwal, 1987.

Results of the pooled data indicated that Brix, sucrose per cent, purity and CCS per cent decreased due to *Pyrilla* infestation, however, the non-sugars increased appreciably (27.3%) in infested canes; this increases production of molasses and concomitantly decreases production of sugar.

As a result of *Pyrilla* infestation, recovery of sugar and *gur* is adversely affected (Kalra, 1983). In the year of epidemic, as much as 50 per cent reduction in sugar and *gur* has been reported (Gupta and Avasthy, 1960; Avasthy, 1973). It is generally observed that concentrated mass does not set well from a badly infested sugarcane and results in poor quality *gur*. According to Gupta (1947; 1948), the production of *gur* is reduced by 2.2 to 4.5 per cent.

Effect of Gurdaspur Borer

Acigona steniellus, commonly known as Gurdaspur borer, is another important pest of sugarcane that causes wholesale damage to the crop. The pest remains active from June to

October and generally infestation varies from 7-35 per cent in different varieties. During epidemic, the damage to the cane may go as high as 40-50 per cent. Singh *et al.* (1957) estimated a loss of 5-15 per cent in cane yield, 17 per cent in total solids, 29 per cent in sucrose and an increase of 84 per cent in glucose. The dramatic increase in glucose content was invariably responsible for poor quality *gur*. The recovery of *gur* is lowered by 74 per cent in the infested canes (Kapoor, 1957).

Effect of Black Bug

In severe attack of black bug *Macropes (Cavelerius) (excavatus)* growth of crop is reduced and the quality of *gur* prepared is adversely affected (Kalra, 1983).

Effect of Pesticides on *Gur* Quality

Studies conducted at Shahjahanpur, in subtropical India, indicated that application of heptachlor (5.6 l (20 EC) in 1685 l of water ha^{-1}) + aretan gamma (6% @ 280 g in 112 l of water ha^{-1}) increased the yield of cane and *gur* per ha over control by 21.4 t and 2.74 t ha^{-1}, respectively (Misra and Singh, 1964).

Nematodes, the hidden enemy of nature, cause substantial damage to sugarcane roots. Although considerable information has been amassed in the literature on the successful control of nematodes by nematicides, the information on nematicides *vis-a-vis gur* quality remains largely obscure. Not many studies of this type have been conducted, though there is an urgent need to take up such studies because of the increasing use of nematicides in sugarcane culture.

Nath *et al.* (1976), for the first time, studied the effect of two commonly used nematicides, *viz.*, Nemagon and Dasanit on the quality of *gur* (Table 12.7).

The quality of *gur* is generally judged by five parameters namely, pol, reducing sugars, colour value, ash content and pore size. Keeping all these attributes in view, Nemagon emulsifiable concentrate (applied at the rate of 60 l ha^{-1} in two split doses) showed marked improvement in quality parameters (Nath *et al.*, 1976). The attributes under other treatments, however, remained *at par* with the control.

Table 12.7: Effect of Nematicides on the Quality of *Gur*

Treatment	Pol (%)	Reducing sugars (%)	Colour value	Ash (%)	Pore space cc/100 g *gur*
Dasanit (25 kg a.i./ha)	79.0	4.33	100	2.82	20.3
Dasanit (12 kg a.i./ha)	79.5	4.27	98.6	2.81	20.6
Nemagon [60 l] (30 + 30 l)/ha]	81.0	4.21	90.9	2.03	25.2
Nemagon (30 l/ha)	79.8	4.27	98.6	2.83	20.3
Control	79.0	4.33	100	2.85	20.0

Source: Nath *et al.*, 1976.

Gur from Diseased Canes

Effect of Diseases (Red Rot and Smut) on Gur and its Quality

Sugarcane suffers from a number of diseases both of stalk and foliage. Although foliage pathogens may not bring appreciable change in the carbohydrate content of the cane, stalk pathogens do cause drastic changes. The magnitude differing with the variety of the cane.

Red Rot

Of a number of seed-piece transmissible diseases, red rot is the most destructive disease and the pathogen causes drastic changes in the carbohydrate content of the juice. As early as in 1929, Rai Bahadur S.M. Hadi reported that when *Paunda (Saccharum officinarum),* badly affected with red rot was crushed immediately after harvest, produced *gur* of a fairly good quality, but when the canes were kept over night and crushed next morning and the juice boiled into *gur,* the final product failed to solidify (Hadi, 1929). Khanna and Chacravarty (1947) reported that besides losses in sucrose, there is increase in the amount of reducing sugars, colloids and ash content in red rot affected canes. A highly deleterious effect of these ingredients was on the clarifying ability of the juice. Mukherjee *et al.* (1980) investigated the production of *gur* from red rot affected canes. They examined the juice samples with regard to pol per cent, purity, glucose per cent and harmful nitrogen in *gur* samples with regard to consistency, colour and taste (Table 12.8). These investigators reported that there was a gradual loss of sucrose in juice with increase in disease intensity. The reverse was true for purity, reducing sugars and nitrogen. Accumulation of non-protein nitrogen not only increases colourisation but also lowers *gur* recovery. The amount of reducing sugars also increased as the disease intensity increased. Increase in reducing sugars, particularly glucose, is undesirable because glucose by virtue of its hygroscopic nature absorbs atmospheric moisture which, in turn, accelerates microbial decomposition of *gur*. Furthermore, high sucrose content and low glucose is a must for inducing crystalline texture in *gur*. Considering that data given in Table 12.7, tolerance limit of red rot affected canes for *gur* making may be kept below 30 per cent disease intensity.

Table 12.8: Effect of Red Rot on Juice and *Gur* Quality of Sugarcane

Treatment (% infection of red rot pathogen)	Juice quality				*Gur* quality
	Pol	Purity (%)	Glucose (%)	Non-protein N (%)	
0	20.07	86.7	0.30	0.02	Hard, crystalline, colour golden, yellow, taste—good
10	15.0	79.2	0.58	0.04	Hard, crystalline, colour—deep red, taste—good
20	14.5	78.5	0.60	0.04	Hard, crystalline, colour—deep red, taste—good
30	13.3	77.7	0.65	0.06	Hard, crystalline, colour—deep red, taste—good
40	12.1	76.7	0.69	0.09	Hard, not quite crystalline, colour—dark chocolate, taste—saltish
50	11.4	72.1	0.73	0.27	Soft, colour—greyish black, taste—saltish
100	7.3	52.0	1.40	0.63	Semisolid, colour—black, taste—very saltish

Source: Mukherjee *et al.* 1980.

Effect of Smut on Gur

Smut is another disease of sugarcane that causes substantial loss to the crop. During epidemic, most of the clumps get affected by this disease. It is now a serious disease in Co 419, Co 740, Co 975, Co 1158, Co 7201, Co 7209, Co 7316, Co 7715 and Co S 510. On the whole a cane with a terminal whip contains about 10 per cent less juice and 3-7 per cent less sucrose than the healthy cane. Sankpal and Nimbalkar (1979) reported that smut infected plants contain more reducing sugars, but less non-reducing and total sugars than the healthy ones. Mukerjee *et al.* (1979) showed the effect of smut on the *gur* quality using smut infected cane of Co 1158 (Table 12.9).

Table 12.9: Effect of Smut Disease on the *Gur* Quality of Sugarcane Variety Co 1158

Treatment	Quality parameter				
	Brix	Pol (%)	Purity (%)	Glucose (%)	Colour value
Healthy	92.55	72.00	77.80	2.36	100
Smutted	84.63	53.60	73.70	4.39	500
CD at 5 %	1.31	1.09	2.69	0.48	54

Source: Mukerjee *et al.*,1979.

The analytical results given in Table 12.9 clearly indicate that *gur* from smutted cane of Co 1158 was significantly inferior to *gur* obtained from healthy canes.

Besides red rot and smut, infection by mycoplasma like organisms (*e.g.* grassy shoot disease), bacteria (e.g. leaf scald) and viruses (e.g. mosaic) generally decreased sucrose content and increased the amount of reducing sugars (Madan *et al.*, 1992, Dhumal and Nimbalkar, 1982; 1983; Agnihorti, 1984). Severe infection by these pathogens appreciably reduced juice extraction percentage per unit cane weight and also adversely affected *gur* production.

Rodents

Rodents are omnipresent pests of sugarcane crop. As the later provides most favourable site for their colonization in terms of shelter and sustained availability of food. Rodents frequent this crop right from planting but their population density gradually increases with the developing crop and attains a peak during ripening phase. Rodents damage being relatively more along irrigation channels, barren land, human habitation, paddy crop, ratoons, thin and lodged canes and in the bunds (Srivastava, 1992). Rodents damage sugarcane crop in following ways:

(i) Nibble growing tissues like buds, young shoots and apical growing part,

(ii) Nibbling roots and stubble while burrowing,

(iii) Nibbling into lower internodes (1-5 above the soil); as a result the next few feet of cane dies and it becomes susceptible to insect pests and as a result reduce sugar yield. Losses in sugar recovery are relatively more due to attack during the late crushing period than in the early part of the season. The overall loss being 12.77 to 24.5 per cent (Srivastava, 1992).

REFERENCES

Agnihotri, V.P. (1984). Changes in juice quality in leaf scald affected sugarcane varieties. *Indian Sugar Crops. J.*, 12: 17-18.

Anonymous (Shahjahanpur , India) (1983). 70 Years of Sugarcane Research (1912-1981), U.P. Council of Sugarcane Research, Shahjahanpur, U.P., pp. 117 + 74 + 59.

Avasthy, P.N (1973). Control of important sugarcane pests. *Seminar on Crop Protection in Sugarcane, Deccan Sugar Technol. Assocn.* Sept., 30 pp, 77-19.

David, H. and A.N. Kalra (1965). A study of sugarcane internode borer *Proceras indicus* Kapur. *Proc. Intern. Soc. Sugar Cane Technol.*, 12: 1444-1453.

Dumal, K.N. and J.D. Nimbalkar (1982). Mineral nutrients in the GSD infected sugarcane leaves of clones of Co 419 and Co 740, *Sugarcane Pathology Newsletter*, No. 28 : 24-32.

Dhura, M.S. and M.L. Agrawal (1987). Effect of *Pyrilla* infestation on juice quality in sugarcane. *Proc. Annl. Conv. S.T.A.I.*, 50 : 161-164.

Gupta, B.D. (1947). *Pyrilla* out break in cane areas of west U.P *Indian Sugar*, 10: 69.

Gupta, B.D. (1948). Control of sugarcane leaf hopper (*Pyrilla* spp.) in the U.P. *Indian Fmg.*, 9: 401-404.

Gupta, B.D. and P.N. Avasthy (1952). Biology of the sugarcane borer, *Emmalocera depressella* Swinh. in the U.P. *J. Agril. and Animal Husbandry, U.P.*, 2 (8): 19-25.

Gupta, B.D. and J. Kulshrestra (1960). Deterioration in the quality and quantity of *gur* and *khandsari* sugar as a result of stalk borer infestation. *Indian J. Sugarcane Research and Devpt.*, 4(2): 64-67.

Hadi, S.M. (1929). The Indian Sugar Industry, Govt. of Bhopal C.I., Published by Khan Bahadur S.M. Hadi, Director of Agriculture, Bhopal State, and printed by F.W. Petch Thacker and Co. Ltd., Bombay, pp. 289 + vi.

Kalra, A.N. (1983). Sugarcane Pests and Their Control, I.C.A.R., New Delhi, pp. 53.

Khanna, K.L. (1957). On the Coccid. *Melanaspis glomerata* Green on sugarcane in Bihar, *Indian J. Sugarcane Res. and Devpt.*, 2(1): 18.

Madan, V.K., S. Solomon and V.P. Agnihotri (1992). *Biochemical Aspects of Sugarcane Diseases*, Indian Institute of Sugarcane Research, Lucknow, India, pp. 90.

Misra, G.N. and J.D. Singh (1964). Improving crop stand and final yield of sugarcane and *gur* through application of chemicals (fungicides and insecticides) at planting time. *Proc. All India Conf. S.R.D.W.I.*, 5: 119-131.

Moholkar, P.R., S.J. Ranadive (1973). Carryover of scale insect (*Melanaspis glomerata*) Green of sugarcane on the quality and recovery of jaggery (Gur). *Proc. Annl. Conv. D.S.T.A.*, 25: 41-44.

Moholkar, P.R., S.J. Ranadive and D.V. Javelkar. (1973). Effect of scale insect (*Melanaspis glomerata* Green) of sugarcane on the quality and recovery of jaggery (*Gur*). *Proc. Annl. Conv. D.S.T.A.*, 25 : 41-44.

Mukherjee, S.R., A. Jha, R.D. Sahi and C.R. Prasad (1980). Investigation on the intensity of red rot disease (*Colletotrichum falcatum*) affecting quality on cane juice and jaggery (*gur*) in the calcareous belt of north Bihar, *Proc. Annl. Conv. S.T.A.I.* 44 : Ag. 99-108.

Mukherjee, S.R., R.P. Nath and M.G. Haider (1979). Effect of smut disease on the jaggery quality of sugarcane. *Indian Phytopath.* 342: 627-628.

Nath, R.P., S.R. Mukerjee, M.H. Asraf and M.G. Haider (1976). Effect of some nematicides on *gur* quality of sugarcane. *Indian Phytopathology*, 29(4): 462-464.

Nigam, L.N. (1960). A preliminary study of the incidence of sugarcane pests and the extent of damage caused by them in North Bihar. *Proc. Indian Sci. Cong.*, 27 (Section B) : 181.

Prabhakar Rao, C.C. Sheshagiri Rao, and S. Venkateswara Rao (1976). Sugarcane scale *Melanaspis glomerata* (Green) and its control measures. *Proc. All India Seminar on Sugarcane Scale Insect, Nidadavolu,* West Godavari, Andhra Pradesh, pp. 68-73.

Prasada Rao, K.K., K. Ramalingaswamy, P.V.L.N. Raju and G. Vishwanathan (1984). Studies on jaggery from hot water treated immature, mature and deteriorated cane. *Maharashtra Sugar,* 9(4): 39.

Ramakrishna Rao, K., Veerbhadra Rao, and M. Radhamohan Rao (1986). Studies on jaggery from hot water treated scale infested cane. *Indian Sugar:* 63-66.

Ramakrishna Rao, K., Veerbhardra Rao, and M. Radhamohan Rao (1991). Studies on jaggery from hot water treated scale infested cane, presented at the *National Seminar on Increasing Productivity, Quality, Storability of Jaggery (Gur),* held at the Indian Institute of Sugarcane Research, Lucknow, organised by Indian Institute of Sugarcane Research, Lucknow and Indian Society of Agricultural Engineers, New Delhi, Nov., 30-Dec., 1, 1991.

Sankpal, S.D. and J.D. Nimbalkar (1979). Effect of smut infection on sucrose metabolism in sugarcane. *Indian Phytopath.* 32 : 306-307.

Shakunthala, V.A. (1991). *Gur* quality and recovery of CoC-671 infested by *Marsipococcus* sp. - a new scale insect in Cauvery delta region, presented at the *National Seminar on Increasing Productivity, Quality, Storability of Jaggery (Gur),* held at the Indian Institute of Sugarcane Research, Lucknow, organised by Indian Institute of Sugarcane Research, Lucknow and Indian Society of Agricultural Engineers, New Delhi, Nov., 30-Dec., 1, 1991.

Singh, H. and B.N. Pandey (1967). Effect of stalk borer incidence on sugar recovery and the quality and quantity of *gur*. *Proc. Annl. Conv. S.T.A.I.,* 35: 329-332.

Singh, O. and K.S. Waraitch (1977). Metabolic changes induced by *Colletotrichum falcatum* Went in sugarcane. *Sugar Cane Path. Newsl.,* No 19 : 7-9.

Smith, N.H. (1959-60). Laboratory studies of cane blanching. *Annl. Rept. Expt. Stn. Hawaiian Sugar Planters Assoc.* (1959-60).

Sreenivasalu, P., B.C. Raju and M.V. Naidu (1985). Carbohydrate metabolism in mosaic virus infected sugarcane (*Saccharium officinarum*). *Indian Bot. Reporter* 4 (2) : 129-133.

Srivastava, D.C. (1992). Production losses due to rodents and control strategies—Sugarcane, in *Rodents in Indian Agriculture,* Vol. I State-of-Art (eds. Prakash, I. and P.K. Ghosh), Science Publishers, Jodhpur, India, pp. 231-248.

13

GUR FROM PLANTS OTHER THAN SUGARCANE

A. PALM *GUR**

Man, through ages, has exploited plants to satisfy his 'sweet tooth'. The common sources, viz. sugar from sugarcane, sugarbeet and sugar maple are well known. The less known are those, that, because of raw material, technological and informational constraints, remain as 'exotics' but have persisted with common sources of sugar. Plams are one such source. They belong to the family *Palmae*. From plams *gur*, syrup and toddy are made.

Although palms find different uses like food, shelter, clothing, timber, fuel, building material, thatch, fibre, paper, starch, oil, wax, dyes and many other materials by mankind (Mc Currach, 1960), its sap is particularly used for producing *neera*, a fresh sap, the sucrose content of which varies from species to species. It is a clear almost transparent, sweet, pleasant smelling and refreshing drink (also known as fresh toddy). Toddy is prepared by fermenting *neera*. This popular liquor of rural areas is a pale frothy liquid having a characteristic aroma. *Gur* (jaggery) is obtained by straining and boiling the juice (sap) till it solidifies. State-of-art of palm *gur* in India up to 1950 has been lucidly presented by Roy (1951). It is a dark, solid substance with characteristic aroma. Sometimes deliquiscent salts are present in it. In wet weather it absorbs moisture and becomes a sticky semisolid mass. Toddy on further fermentation and distillation gives an alcoholic beverage known as *arrack*, also called palm wine. On acetic fermentation, toddy gives vinegar or *sirka*.

---

*Contribution from: K. Sengupta, S.K. Das and C. Sen, Bidhan Chandra Krishi Vishwavidyalaya, Kalyani, West Bengal, India.

Because of low production and regional distribution of source, the technology of palm *gur* production has remained traditional but akin to that used for sugar maple (Kulkarni, 1971). Various aspects of palm *gur*, viz., source plant, tapping of juice, and its processing to yield *gur* are given below.

The Source: Family *Palmae* (*Arecaceae*)

Several species belonging to the monocotyledonous family *Palmae (Arecaceae)* have been exploited for tapping the sugar rich sap (Table 13.1).

Table 13.1: Palms and Their Parts Containing Sugar which Serve as a Source of *Gur*

Palm	Part(s) containing sugar
Palmyra palm (*Borassus flabellifer* L.) (Syn. *B. flabelliformis* Murr.)	Spadix or fruit
Wild date palm (*Phoenix sylvestris* Roxb.) (Syn. *Elate sylvestris* L.)	Shoot-tips (tender upper portion of the stem) or spadix
Coconut palm (*Cocos nucifera* L.)	Crown base
Sugar palm/Gomuti palm (*Arenga pinnata* (Wurmb. Merril (Syn. *A. saccharifera* Lab.) *Gomotus saccharifera* Spreng.)	Peduncle (stalk of inflorescence)
Fish tail palm/Toddy palm (*Caryota urens* L.)	Inflorescence
Nipa palm (*Nypa fruticans* Wurmb.)	Peduncle
Talipot palm (*Corypha umbraculifera* L.)	Inflorescence
Buri palm (*Corypha elata* Roxb.)	Inflorescence
Sago palm (*Metroxylon sagu* Rottb.) (Syn. *M. sagus* Rottb.)	Trunk

Geographical Distribution

The broad geographical distribution of palms, exploited for *gur* production, is shown in Figs. 13.1 and 13.2.

Palmyra Palm (Bengali: *Tai,* Hindi: *Tar*)

It occurs throughtout the drier areas of tropical Asia and Africa and is abundant in India. It is distributed mostly in Myanmar, Sri Lanka and Malayasia. In India, the palm is found in dry

Fig. 13.1: Global Distribution of *Gur* Yielding Palms
Palmyra palm +; Wild date palm ○; Coconut palm △; sugar palm ●; Fish tail palm ▲; Tail pot palm □; Nipa palm ■

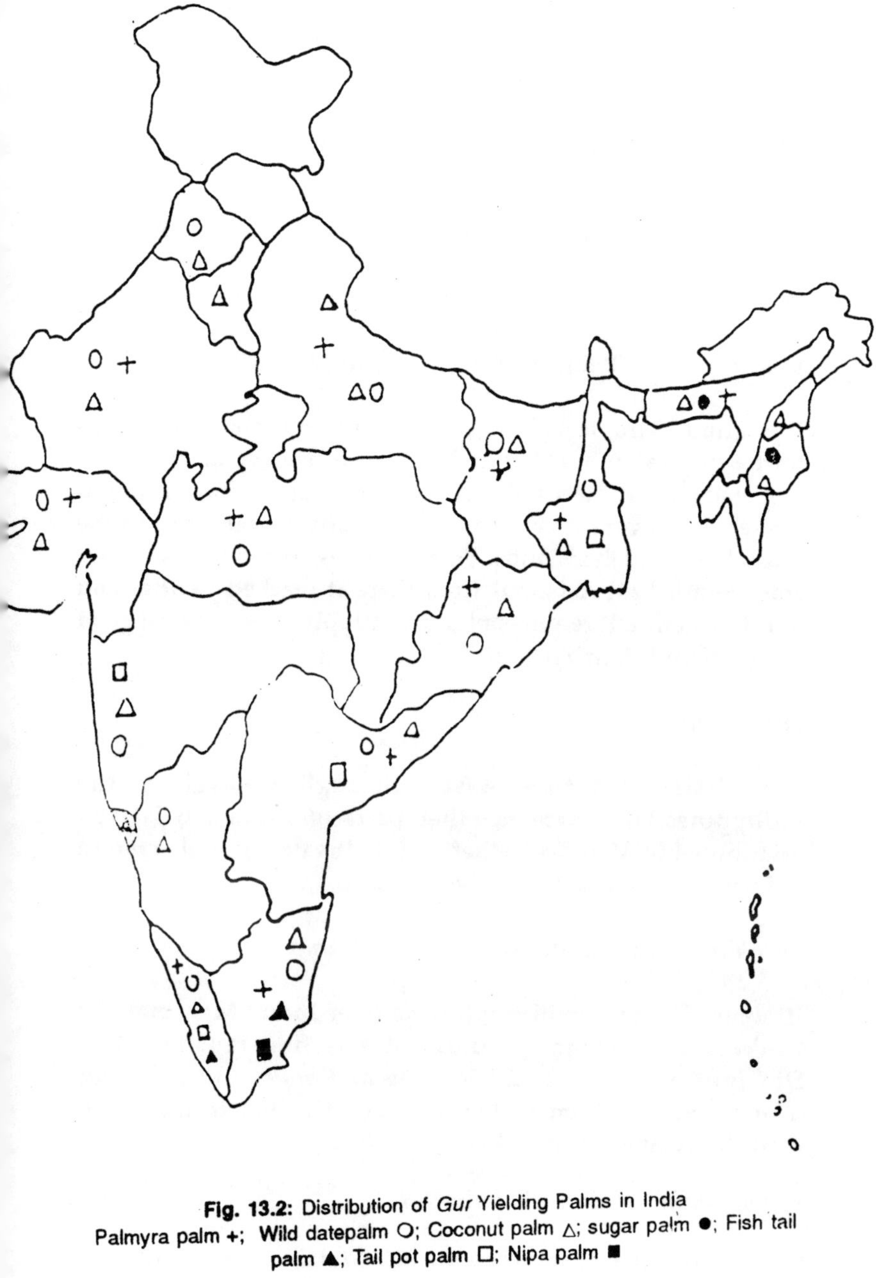

Fig. 13.2: Distribution of *Gur* Yielding Palms in India
Palmyra palm +; Wild datepalm ○; Coconut palm △; sugar palm ●; Fish tail palm ▲; Tail pot palm □; Nipa palm ■

regions along the coastal areas of the peninsula and in West Bengal and Bihar. These palms grow abundantly in sandy plains just above the sea level having annual rainfall of 620 to 650 mm.

Wild Date Palm (Bengali: *Khajur*, Sanskrit: *Kurjura)*

It grows mostly in India, Bangladesh, Pakistan, Sri Lanka and Myanmar. It is distributed throughout the tropical belt and occurs in almost all the States of India except hilly areas. It is often confused with *P. acaulis* Roxb., another species.

Coconut Palm (Bengali: *Narkel*, Hindi: *Nariyal)*

It is a humid tropical plant. Its distribution is almost confined to the tropical zones that include South East Asia, India, Sri Lanka and pacific territories of West Indies. In India, it is found scattered all over the country but is mostly concentrated in the coastal belt and the islands. This palm has spread to East Africa from South East Asia and from there it reached Central and South America. It requires plentiful supply of water with good underground drainage.

Sugar Palm

It is distributed from Assam through Malaysia to the Philippines. It is rare in other parts of India but is well distributed in Malayasia where it is cultivated. It is also found wild in Myanmar and southern China.

Fish Tail Palm (Hindi: *Mari*, Greek: *Karyon)*

It occurs in the humid tropical parts of India, Myanmar, Sri Lanka and Thailand. In India, it was first reported from Dinajpur and known as *Bon (Bura) Khajur* and is found distributed in Assam in the east and Malabar in the south west. In Telangana, it is known as *Jeroogoo.*

Talipot Palm (Hindi: *Bhajarbattu)*

It originated mainly in Sri Lanka and Malabar coast. However, it grows wild in Bengal, Andaman and Nicobar islands, Malabar, drier regions of Malaysia and northern Australia.

Nipa Palm (Hindi: *Golphal* or *Gulga*)

It occurs wild in Sri Lanka, India, Philippines, Carolines, Solomon Islands and Queensland (Australia).

Buri Palm (Hindi: *Bajoor*)

It is a source of sugar in Philippines. In Amazonia, the *assai* palm produces berreis from which refreshing drinks and beverages are made. Trees of *sago palm* grow wild in Coromondal coast, Malabar, Maharashtra and Madhya Pradesh.

Taxonomic Features of Source Plants

Palms belong to the family Palmae. More than 2600 known species are economically important. Palms are chiefly tropical and subtropical trees, shrubs and vines usually with a tall columnar trunk. The trunk is crowned by a very large flattened fan shaped or feather shaped leaves with stout leaf sheaths often having prickly petioles. The persistent bases of which frequently enclose the trunk (*Encl. Brit. (Micropaedia)* VII: 699, 1978). Soviet biogeographer N.I. Vavilov considered that sugar palm and coconut palm originated in *Indonesia-Indochina* genetic centre occupying vast areas of South-east Asia. This area happens to be also the centre of origin of vegetatively propagated crops like sugarcane and banana (*Encl. Brit. (Macropaedia)* 5: 938, 1978).

The taxonomic features of palms including the species of sugar yielding varieties are well described in classical works of Griffith (1850), Blatter (1926), Bailey (1930), Loomis (1949), Hawkes (1950-52) and more recently in *Palms of the World* by Mc Currach (1960). The salient features of various palms are presented below:

Trees with unbranched erect, stout stem and a crown of pinnately or palmately sected leaves are diagnostic characters of the family (Mitra *et al.*, 1990). Stems are covered with leaf scars of persistent leaf stalks. Inflorescence is simple or compound (paniculate spadix) subtended by one or more usually large bracts (spathes) of various forms. Flowers are very small, almost sessile, unisexual, actinomorphic. General floral formulae are:

Male flower: ♂ $P_{6\ (3+3)}\ A_{6\ (3+3)}$ or $_{3}$
Female flower: ♀ $P_{6\ (3+3)}\ G_{(3)}$

Fruits are usually berry or drupe with endocarp attached to the seed coat (Fig. 13.3).

The chief palms utilised for *gur* production are palmyra palm, the wild date palm and the coconut palm. The juice/sap of these species of palms is a major source of commercial *gur* in some tropical countries. Salient taxonomic characteristics of such palms are presented in the following paragraphs:

Palmyra Palm

Aged tree atttains a height of 10-18 m and a girth of 1-2 m and bears a terminal crown of 20-40 large fan like leaves, 90-150 cm in width. The leaf stalks are 60-110 cm long, strong and fibrous. The stem is dark in colour and its hard outer portion mainly composed of stiff longitudinal fibres, while the pith (central portion) is soft and starchy. They are dioecious. The fruit is large, fibrous, containing usually 3 nuts each containing a seed (Fig. 13.4).

Wild Date Palm

It is a pinnate-leafed palm with pointed leaflets. Plants are dioecious with a trunk 20-30 m in height and a terminal crown of 80-120 leaves. Inflorescence axillary, much-branched, enclosed in a deciduous spathe. Fruits are cylindrical, yellow to reddish brown; ripening takes 5-6 months. Seeds are cylindrical, deeply grooved with very hard endosperm (Fig. 13.4).

Coconut Palm

Coconut palm is a tall tree, 12-30 m tall; the trunk is marked by prominent ring-like leaf scars. The main stem is unbranched,

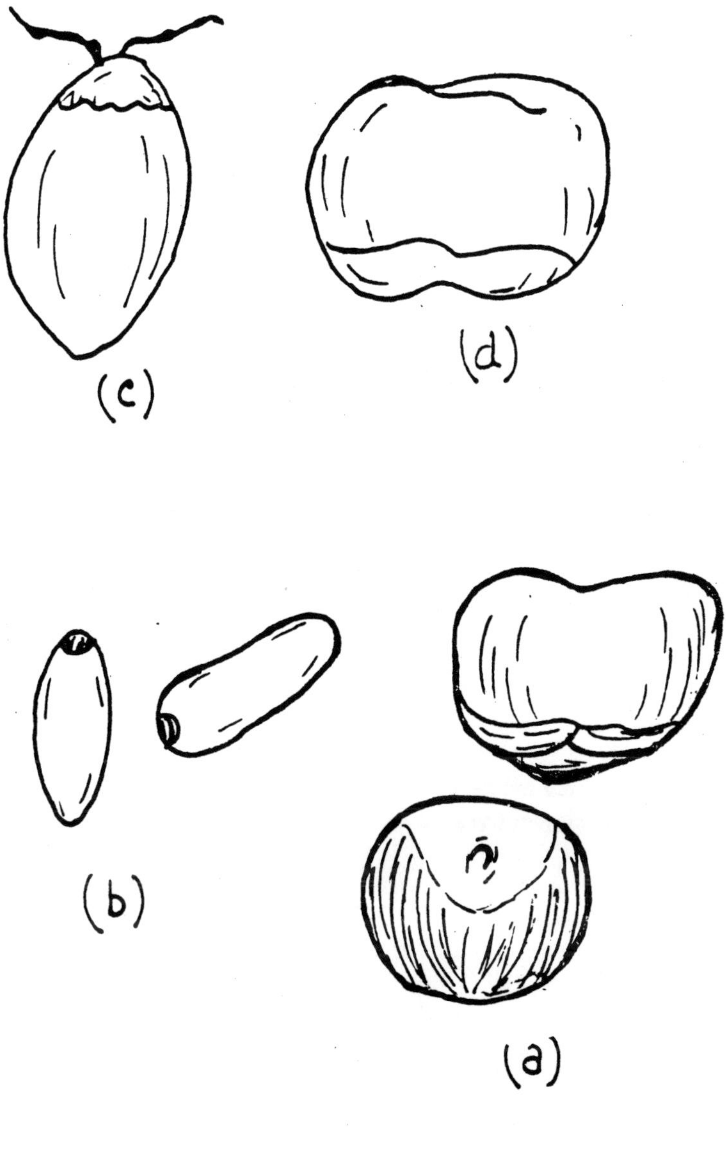

Fig. 13.3: The Morphological Features of Fruits of Important *Gur* Yielding Palms: (a) Palmyra palm, (b) Wild date palm, (c) Coconut palm, (d) Sugar palm

with a swollen base surrounded by a mass of adventitious roots and is crowned by 20-30 large, peripinnate leaves. The spadix (inflorescence) is produced in the leaf axil. The palm flowers all the year round. Mature fruit is a fibrous drupe (usually ovoid) (Patel, 1938) (Fig. 13.4).

Sugar Palm

It is a monocarpic monoecious feather palm, 6-12 m tall; large pinnatisect leaves form a diffuse crown. Flowering begins (when the palm is 7-10 years old) at the top of the trunk and continues downwards; inflorescence has a stout peduncle; male and female flowers occur on separate peduncles; first 2-5 inflorescences are usually female, then more numerous male spadices come out. The fruits are obovoid, globose, 2-3 seeded with semi-pulpy kernel (Fig. 13.4).

Fish Tail Palm

A single-stemmed monocarpic palm, 10-20 m tall with prominent internodes; leaves are bipinnate, about 6 m long, mature pinnae end in terminal leaflet; pinnules like eroded fish tails, wedge-shaped with no distinct midrib but veins are prominent, tips broadly toothed. Inflorescence subtended by several bracts; flowers arranged in groups of 3 with single female between 2 males; male flowers open first. Fruits globular, fleshy without woody stone (Fig. 13.5).

Nipa Palm

Leaves are pinnate, unarmed with air cavities in buoyant leaf bases (Purseglove, 1975). Tree is monoecious. Inflorescence with globose terminal head of female flowers surrounded by numerous short spikes of male flowers. Fruits are obovoid, angled, ribbed with smooth exocarp, fibrous mesocarp and firm endocarp, one seeded (Fig. 13.5).

Talipot Palm

It is a monocarpic fan palm, very tall; the trunk is covered with persistent leaf bases. Leaves are orbicular, petioles are long

Fig. 13.4: Features of *Gur*-Yielding Palms : Top left : *Borassus flabellifer*—a young plant, Top right: *Phoenix sylvestris;* Bottom left: *Cocos nucifera*—a dwarf variety; inset shows close up of fruits; Bottom right: *Arenga pinnata*

Fig. 13.5: Features of *Gur* Yielding Palms : Top left : *Caryota urens*; Top right : *Nypa fruticans*; Bottom left : *Caryopha umbraculifera;* Bottom right : *Caryopha elata*

with spiny margins and thick broad bases. The vast terminal inflorescence is the largest in the plant kingdom and is produced after 25-30 years; inflorescence is much-branched, pyramidal (Fig. 13.5).

Sago Palm

Trees are monoecious, monocarpic with large pinnate leaves and basal suckers. Each trunk produces a vast terminal inflorescence at the age of 10-15 years. The flowers are borne spirally in pairs, of which one is male and the other hermaphrodite but acts a female. Fruits are globose, borne is whorls with little stone and a single seed.

Characteristics of various palms yielding palm *gur* in India are given in Table 13.2.

Tapping the Sap

The first prerequisite for preparing *gur* is to have a good efficient system of tapping without over exploiting the genetic capacity of the plant to produce it, and thus sustaining its production over the years. The method of tapping is essentially similar for most of the plants but differs in details commensurate with their morphology, optimum duration of sap release and part of the plant to be tapped. Some of the tools used in tapping sap are given in Fig. 13.4. These are described for each possible source plant.

Palmyra palm

Tapping normally begins when the palm is 25-30 years old and may continue for 30 years. Palms growing in the coastal sandy soils are ready for tapping at the age of 15 years.

All inflorescence of both sexes may be tapped, but the male tree yields more than the female tree. In case of male trees, the flowering shoots having finger like growths (on the principal stalk) are bruised by rubbing them between wooden sticks. After a few days, these are scrapped with a tapping knife and their tips are shaved off daily. When the juice begins to ooze out (within a week after bruising), several fingers are tied together and bent into a pot for collection. The female trees are tapped when the nuts are still very small, and only the tips of the spadix are squeezed and beaten. They are then pared off just like male trees.

Table 13.2: Characteristics of Various Palms Yielding Palm *Gur*

Particular	Date palm	Palmyra palm	Coconut palm	Sago palm
Life of tree (years)	50-60	90-120	60-80	35-45
Commencement of tapping from	8th year	15th year	8th year	10th year
Tappable part of the tree	Crown base	Spadix or fruit	Spadix	Spadix
Tapping season	Oct./May	Jan./July	Throughout year	Throughout year
Overall tapping period (years)	25-40	70-95	25-35	15-20
Tapping period in a year (months)	4-6	Male 2; Female 4-6	4	6
Average yield of juice/tree/day	2.35 kg	2.62 kg	1.88 kg	9.90 kg
Seasonal yield *gur* (kg)/tree :				
(a) Maximum	23.5	70.5	37.6	185.0
(b) Minimum	7.1	18.8	28.2	94.0
Average sucrose % in juice	10.0	12.0	14.0	10.0
No. of trees tapped by one tapper/day	20-40	10-80	10-30	5-10
No. of trees/ha	1235	1235	198	247
Mode of tapping	One days tapping followed by 3 days rest	Thrice a day	Twice or thrice a day	Twice or thrice a day
Yield of *gur* % juice (on weight basis)	10-15	10-16	12-18	8-11

Source: Indian Sugar Manual, 1956, pp. 256.

Wild Date Palm

From the age of 7-10 years, when the trunk of the trees are likely to be about four feet high, the plants begin to yield juice and continue productivity for 20-30 years.

Tapping is done during the cold months—November through February. As the temperature rises above 20°C during summer, there is deterioration, both in the quality and quantity

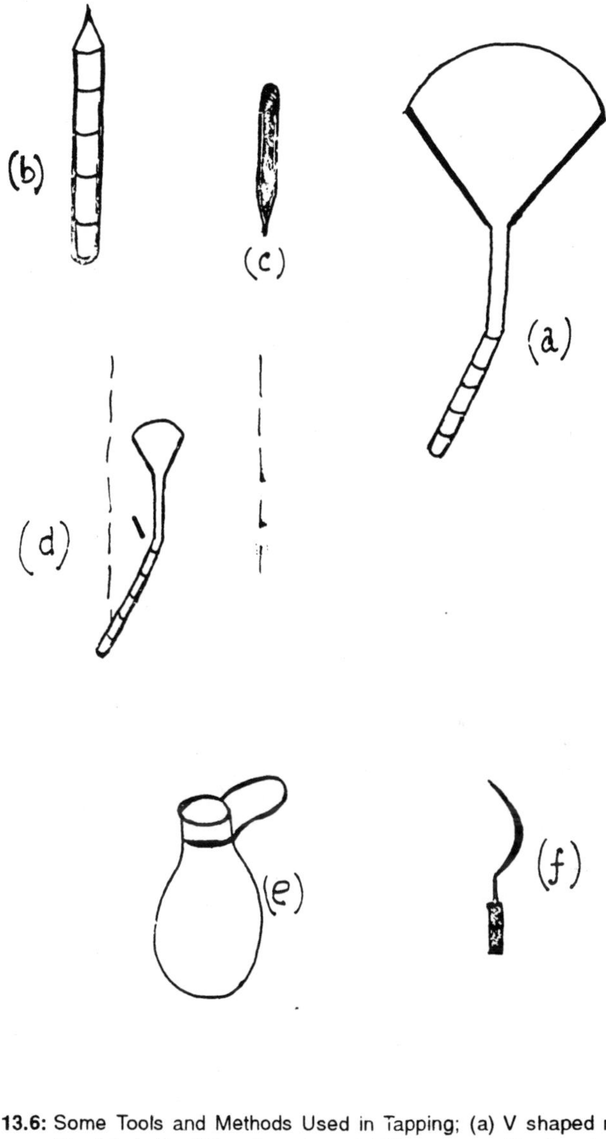

Fig. 13.6: Some Tools and Methods Used in Tapping; (a) V shaped notch; (b) The thin half-split bamboo piece; (c) The peg used for hanging the pot; (d) V notch for bamboo tongue and peg; (e) Mud pot used for collection of sap; (f) Knife used for cutting

of the juice and the tapping is discontinued. The first operation step consists of cutting off the lower leaves along with the leaf base on one half of the crown, leaving only the vertically disposed younger leaves uninjured. In this way, the tender stem on this particular side is exposed. No operation is effected on the tree during the following 6-7 days. After this interval, a V-shaped layer about 15 to 20 cm broad, 30 cm long and nearly 1 cm thick is sliced off (known as *jiran* cut) in thin layers from the freshly exposed surface of the stem. In different regions, there are variations in the size of tapping as well as in the shape of tapping knife and juice collection vessels used. The sap trickles down the exposed surface and flows downward reaching the narrowest point of 'V' where a thin half-split bamboo piece is fixed with its free end jutting out as a tongue through which the sap flows into a small pitcher (mud pot) hung in position. During the first night of this tapping, a large quantity of high density (strong) juice exudes which is known as *jiran-kat* (Fig. 13.7). The second evening the cut surface is gently scrapped to freshen the wound (known as *majher-kat* and to ensure continuous flow of sap. By quantity and quality, the juice collected during the second night is inferior to *jiran* and is known as *majher-kat* (*do-kat*) (Fig. 13.7). The third evening, the cut surface is merely rubbed or slightly pared off (known as *Jharra* cut) and the juice collected during the third night, known as *jhara,* is the poorest in quantity and quality. After three days (nights) of collection, the tree is given a rest for the next three complete days, after which the same process is repeated. After every six days, a new cut is made over the previous one, the stem gets more and more hewed into as the season progresses.

Coconut Palm

Flowering begins when the trees are 4-5 years old; the palm generally starts bearing at the age of 5 or 7 years after planting and the stabilised yield is obtained from 10th year onwards for 50 to 60 years (Menon and Pandalai, 1960).

The methods of tapping of coconut palm differ in detail in different locations (Nathanael, 1956; Marsden and Skilton, 1931). However, the basic principle remains the same. In the

Fig. 13.7: Tapping Processes of Wild Date Palm; (a) A typical plant (3-year old); (b) Preparing the plant for tapping—the *jiran* cut, along with placement of a bamboo tongue for facilitating collection of sap; (c) The second cut - *majher* cut, to freshen the wound; (d) Hanging a pitcher for collection of sap.

western parts of India, the inflorescence suitable for tapping is first 'trained' (prepared). The training consists of gentle uniform tapping (beating) of the entire surface of unopened flowers (spathe) twice a day, in the morning and in the evening. Then, three-forth of the lower length of the spathe is tightly wound round with a strong string, commonly the fibrous chord of the coconut leaf petiole or coir, to prevent the spathe from opening. In some cases, where there is fear of the spathe bursting, tying precedes tapping (beating). After a week's beating, about 7 to 10 cm of the tip of the spathe is cut off. In some places, this naked surface is tapped with the handle of the tapping knife. This process is continued in the morning and evening for six to seven days according to requirement. By this time, the surface appears moist due to the exudation of juice. The early and proper flow of juice depend mainly up on the skill of the worker. Special care is taken not to reduce the flower buds inside the spathe to a pulp in which case the spadix becomes useless. In some places, a 2 to 7 cm wide ring of the spathe from the tip is removed carefully without causing any injury to the spikes. This fresh portion is tied up with tender leaflets. A thin slice is pared off the end of the spathe twice, both in the morning and evening or once a day. The naked surface is often protected with a cap made of a leaf stipule.

When the above operations are continued for a few days, the juice begins to trickle in the form of drops. When the flow of juice starts, the paring of the tip is done thrice a day. A receptacle, such as an earthen pct (pitcher) specially made for the purpose is hung from the cut surface for the juice to collect in it. The interval from commencement of tapping to the trickling of juice depends on the skill of the tapper, the method of tapping adopted, the seasonal or weather conditions prevailing and the nature and age of the tree. This period usually varies from 10-25 days.

Training the inflorescence is a long process in some places. The unopened inflorescences are selected and are bent downwards slowly and gently, this operation being repeated several times in a day for about one to two weeks. By doing this, the entire inflorescence droops and the cut surface practically faces downwards. This avoids the toddy entering into the inflorescence. Further, it enables the use of cylindrical receptacles like bamboo or bottles to collect the juice. In some places, specially designed mud pots are used as receptacles and a slit is made in the lower part of the spathe, thus avoiding the tedious process involved in the training of the spadix. Toddy is usually collected both in the morning and evening. If the trees are tapped for sweet juice, it is collected invariably in the morning. However, in some places, collection is done twice a day. In the tapping operations on a spadix, after the flow of juice has commenced, shaving a thin slice off the spadix afresh may be done twice or thrice a day according to the requirements. The cut surface is covered with a thin layer of fine clay. Menon and Pandalai (1960) reported from Tamil Nadu that in some places slicing is followed by washing and rubbing the cut with the paste of the leaves of *Achil (Vittil)* tree (*Aporosa lindleyana* (Wt.) Baill). Any mucilaginous leaf can be used for this purpose. This prevents the rapid healing or drying of the cut surface.

In Malayasia, a single coconut palm (cultivar-Mawa palm) can produce 40 kg sugar per year. Annual sap production is 300 l/tree based on tapping of only one spathe at any particular time. Total sap production is 45 l per spathe for a tapping period of 48 days (Omran, 1988).

Sugar Palm

The palm matures in about 5-8 years, when the floral axis is produced. During flowering, starch from the stem is transformed into sugar and is transported to the growing axis of the inflorescence. After the emergence of floral axis the young male spadix is tapped and the juice is collected.

Tapping is done by beating male floral axis with a wooden stick for 3-4 days to ensure the flow of sap into the tapped (wounded) portion. Then a small cut (notch) is made in the peduncle by chipping it. The exudate is collected in the earthen pots previously heated and smoked, or smeared with lime paste or lime water to partially sterilize it. The juice is collected in this manner, day after day, by making fresh chips in the same peduncle. Collection of juice is continued for 6-80 days.

Nipa Palm

Sap is collected when the fruit is beginning to form. Usually the palm flowers in 3-4 years. The palm is not tapped until the second flowering (at about 5 years age) and several inflorescences on each plant may be tapped at one time. Tapping may continue for 2-3 months. Pattern of extraction is same as mentioned for sugar palm.

Sago Palm

In sago palm, spathes are long like elephant trunks. These require a great deal of sqeezing and ruffing for more than a month before juice begins to ooze. The tip is well cut and protected against insects when the sap begins to ooze. It is collected thrice a day in mud pots smeared with lime paste to prevent fermentation.

Lontar Palm

In the eastern Indonesian Island of Sabu, there is a drought resistant *lontar* palm (*Borassus sundaicus*). The tree is tapped twice daily during the flowering stage at the end of dry season, when food stocks are the lowest. The sugary juice provides two meals a day for 4 months and surplus is boiled to a molasses-like syrup which can be stored for several years and some of it is even fed to pigs. This tree also provides material for making houses, buckets, ropes, paper and even contraceptive pills (Bowles, 1990).

Processing Sap to *Gur*

The subsequent processing of the juice (sap) to *gur* is essentially similar and consists of concentrating and solidification in a convenient mould for attractive and easy marketing. But the details of approach differ and are presented briefly for each source.

Palmyra Gur

After collection, the juice is strained into the boiling pot, leaving behind the sediment of lime paste. Some 15 to 20 minutes after the commencement of boiling, a whtie scum rises to the surface which is skimmed off (removed with a ladle) and a few castor beans are crushed and put into allay frothing. *Gur* sets in about 15-20 minutes and amounts to 15-18 per cent of the sap. According to Samba Murthy and Subrahmanayam (1989) the quality of *gur* can be improved by precipitating lime with carbon dioxide gas or citric acid or unripe tamarind fruits, before boiling the juice.

Studies during 1933-34 at the Regional Sugarcane Research Station, Anakapalle (Andhra Pradesh) indicated that the juice of male trees gave higher *gur* recovery (11.7%) than the juice from female trees (10.0%) (Anon., (Anakapalle, India), 1989).

Date Palm Gur

Freshly obtained sap of this palm is clear as water, very sweet, containing 9 to 10 per cent sucrose. Accroding to Davis (1977), application of lime paste in the receiving containers delays fermentation of the juice. When this sweet juice is boiled, a thick syrup is obtained which is consumed in various ways. Sufficiently boiled down juice solidifies into the *gur* which is moulded in suitable blocks (Fig. 13.8).

The *neera* (sweet, clear juice) collected from different fields is poured into a shallow but big aluminium or galvanised iron tray after straining (through plastic or wire net). Then the tray is heated. When boiling starts the scum that comes up to the surface is removed twice with a perforated ladle. The juice is stirred at intervals for rapid evaporation. When the juice turns into a viscous thick mass, heat is controlled (low heat). To test whether the juice has reached the *gur* stage, a little quantity of the viscous mass is taken in a ladle and allowed to drain downwards slowly. If it has reached the right consistency, it forms droplets while falling. The tray is then removed from the oven and placed on the ground. Now, with the help of a long ladle the viscous mass is rubbed against the tray till fine crystals are formed. The rubbing is a must for the preparation of *Patali gur*. The mass is then poured into suitable containers (preferably made of stones or metals). The *gur* so prepared is known as *Patali* or *Labat gur* which is hard, golden in colour,

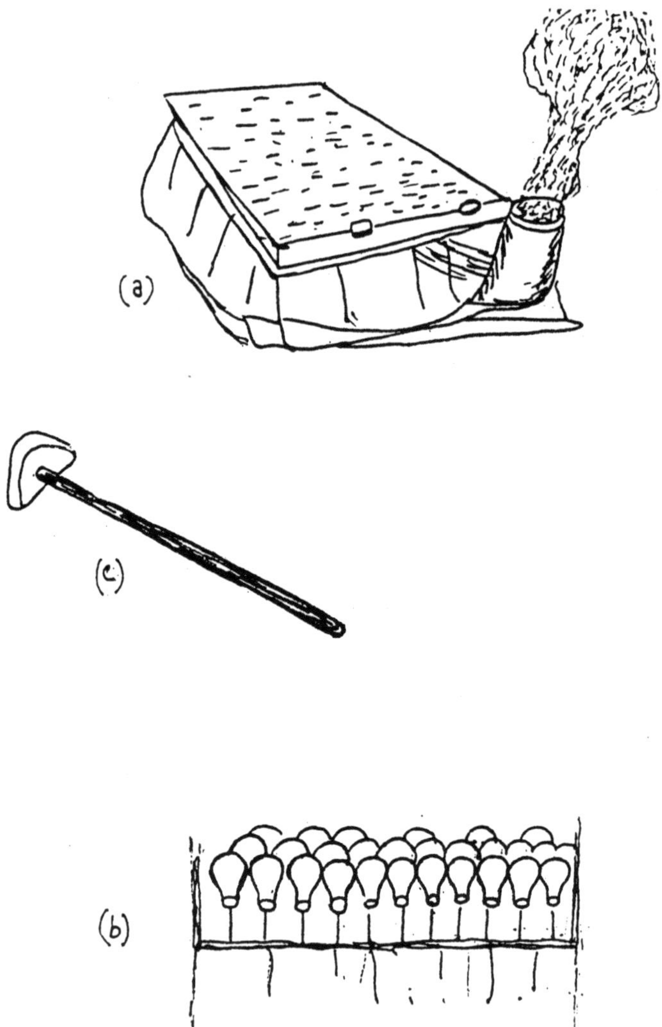

Fig. 13.8: Tools and Techniques Used in Concentrating Sap into *Gur*, (a) Boiling of juice for the preparation of *gur*; (b) Sterilization of pots by heat (fire); (c) Ladle used for rubbing the thick mass.

Fig. 13.9: Processing of Wild Date Palm to *Gur*, (a) The first phase of boiling in metal tray showing the scum floating up; (b) The scum is removed and boiling continued (On the right corner the pots are being partially sterilised by heat).

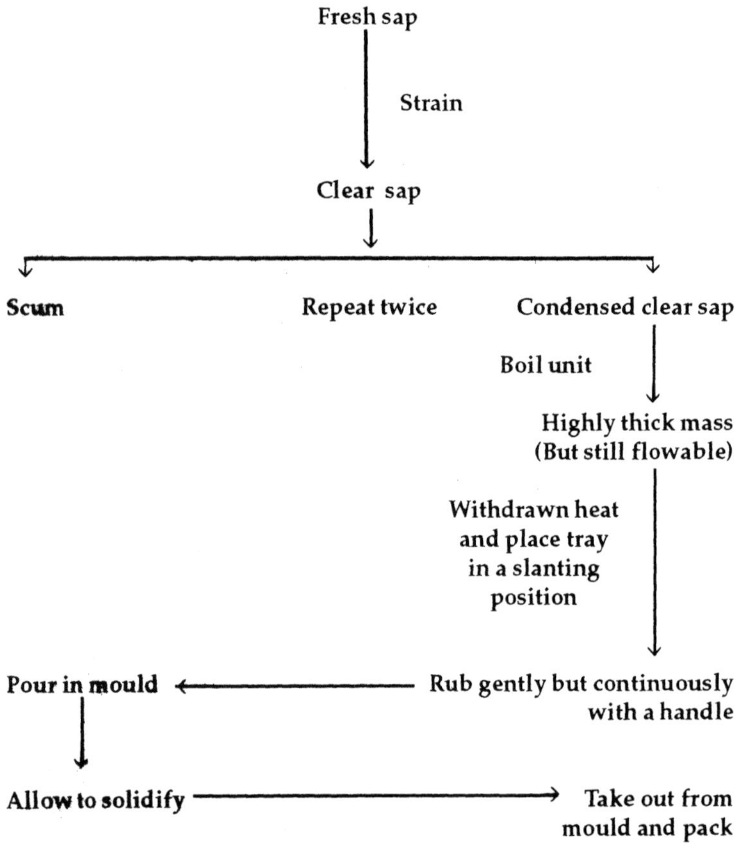

Fig. 13.10: Flow Diagram showing Steps Involved in Preparation of *Khajur Gur* from the Sap of Wild Date Palm

fine, crystalline and a delicacy *par excellence* in West Bengal (Fig. 13.9). A flow diagram showing step involved in preparation of *gur* from the sap of wild palm is given in Fig. 13.10.

Coconut Gur

Fresh sap, the *neera,* when boiled at 118 to 120°C and allowed to cool, solidifies. The solid mass is known as *coconut gur.* The crude way of making *gur* is to strain the *neera* and boil it in tall mud pots till it becomes viscous. The viscous mass is poured into half cut coconut shells or other moulds where it solidifies. The *gur* made in this way is dark in colour and has a caramelised taste. This is due to the presence of excessive lime and lack of uniform heating whereby some of the *gur* on the sides of the pot gets charred. Several improvements over this method have been devised. The procedure now adopted for producing *gur* is as follows:

A shallow but big galvanised iron pan is used instead of a mud pot. Similarly an improved furnace is used to economise fuel consumption. The *neera* (sweet toddy) is first strained through a fine cloth or wire mesh. Then the boiling pan is smeared with any vegetable oil and slightly heated. The oiling of the pan checks caramelisation and lengthens the life of the pan. The strained juice is poured into the pan and slightly heated. Menon and Pandalai (1960) reported that super-phosphate in small quantities is to be added at this stage to bring the pH of the juice to about 7. This precipitates excess lime. The *neera* is then heated to about 70°C. After that, the juice is filtered through a piece of coarse cotton cloth. The filtered juice is again boiled in the pan. When the boiling starts, the scum comes on the surface and is skimmed off with a perforated ladle. The juice is stirred frequently to facilitate rapid evaporation. Light fuel is advantageously used at this stage. When the juice turns into a viscous mass, heat is controlled (reduced). To test whether the juice has reached the *gur* stage, a little quantity of the mass is put in a cup of cold water. If it has reached the right consistency, it turns solid when rolled between the fingers. The viscous mass is then poured into suitable metal or wooden moulds. After allowing the *gur* to set for some time, it is removed from the moulds and suitably packed. The *gur* so prepared, is hard, crystalline and golden yellow in colour. Coconut *gur* is a widely used sweetening agent in West Bengal, has a good flavour and is used in the preparation of several dishes. The *gur* is considered superior to cane *gur*.

The Yield of Palm Gur

The yield obviously depends on sugar content of the juice and the efficiency of extraction process (Table 13.3).

Table 13.3: Yield of Palm Sap from Different Palm Species

Palm	Average daily yield/tree (in l)	Average yield per season/ tree (in l)	Sucrose content in juice/sap (%)
Palmyra palm	2.0 - 2.5	170 - 185	11.5 - 12.00
Wild date palm	4.0 - 5.0	180 - 200	9.0 - 10.0
Coconut palm	1.5 - 2.2	240 - 290	12.0 - 16.0
Sugar palm	3.5 - 5.5	80 - 120	14.5 - 15.0
Fish tail palm	5.0 - 15.0*	350 - 450	12.0 - 15.0
Nipa palm	10.0 - 25.0	380 - 480	15.0 - 17.0

* For a single inflorescence. It is 20 - 30 l when several inflorescences are tapped.

Palmyra Palm

The daily yield of sap from a palm is on an aveage about 2.25 l and the total yield per season is normally 180 l. Corner (1966) reported that an inflorescence yields 2 l per day and a large tree with several inflorescences in tapping will give 20 l per day. A single palm is estimated to produce 1,20,000 l during its tapping life. The yields of sap is higher during the cold months than in the summer.

Wild Date Palm

A tree can yield about 4-5 l (5 kg) *neera* in 24 hours, although exceptionally good ones can produce as much as 10 kg. During the four months of tapping (November to February), a palm yields approximately 180-200 l (200-250 kg) of *neera*.

Yields of juice and *gur* depend on several factors, such as age of the tree, season, soil condition, land situation and number of cuts (serial number). A typical case in point of juice and *gur* yield from Bankura district in West Bengal is presented in Table 13.4.

Coconut Palm

The average daily yield per palm is about two l, the yield per spathe varies from 15 to 18 l and an average palm gives 270 l during eight months of tapping though there is considerable variation in the yield of juice from day to day, season to season,

spadix to spadix and tree to tree. Sap always flows more readily in the night than during the day. During cloudy weather, the flow of sap increases but its consistency decreases. As the dry season continues, the flow of the sap gradually peters of, but consistency increases in dry weather.

Table 13.4: Juice Yield of Wild Date Palm in Different Months in Bankura District of West Bengal

Month	Normal atmospheric temperature (□C)		Average daily yield/plant (kg)	Juice: *gur* ratio	Average daily yield of *gur* (kg)
	Max.	Min.			
November	30.5	16.7	6 - 8	8 : 1	0.75 - 1.00
December	27.1	12.9	6 - 8	6 : 1	1.00 - 1.33
January	26.8	11.6	5 - 6	6 : 1	0.83 - 1.00
February	31.5	14.3	4 - 5	10 : 1	0.40 - 0.50

Source: Misra (personal communication).

Sugar Palm

Female spadices do not yield any sugar; male tree on an average yields about 4 l of juice per day and about 400 g *gur* is obtained from this juice.

Shelf Life of Sap and the Gur

Preservation of juice (*neera*) is a difficult job. When *neera* is employed for the preparation of *gur*, it is usually collected during the night or in the early morning, when the atmospheric temperature is comparatively low. Clean containers (usually earthen pots) are used for the collection. These pots are sterilised previously by heat (burning) and smoked. Sometimes the pots are smeared with slaked lime. Even with all these precautions, on an average, the fresh, clear toddy (*neera*) can be preserved for 10-12 hours only. Juice (sap) remains unfermented for this period if collected carefully in treated containers. As usually collected, unless special care is taken, it rapidly ferments owing to the action of yeast-like microorganisms present. It was found that addition of 0.2 per cent of benzoic acid is necessary for complete inhibition of both alcoholic and acetic fermentations

of coconut *neera* (Anon., 1940), but this quantity of benzoic acid is toxic to human beings and, therefore, is not recommended. Walawalkar (1950) reported that sweet juice can be preserved by the addition of 10-60 ppm of sulphanilamide.

On the other hand, once made into *gur* the shelf life is good when kept under good dry and airy conditions. Moulds grow if stored under uncongenial conditions.

Special Types of *Gur* and Syrup Obtained from Palms

Jhola Gur

From the sap collected during the day *jhola* or liquid *gur* is made in the evening. If weather is cloudy, it is not possible to manufacture solid *gur*. On such days, from the sap collected in the night only *jhola gur* is made. Boiling of the sap is done in the same way as for the solid *gur* but is is not concentrated to the same extent as for solid *gur*. The semi solid mass is then kept in the earthen pots so as to crystallise. *Jhola gur* is sold in the market at a relatively lower price.

Talmisri or Palm Candy

Raw juice is boiled down as before but not to such a high degree of concentration as for preparing *gur*. The syrup is then poured into covered pots and buried underground for some months and during this period candy crystals are formed. This is considered a delicacy and known as *talmisri*. It is known for its medicinal properties and used in cough and pulmonary infections. *Talmisri* solution is administered to children in summer season to avoid heat.

Neera Prash

Neera prash is a soft beverage prepared from palm molasses. The molasses is diluted with water to 60° Brix. It is then boiled and clarified with a little milk acidified with 0.06 per cent citric acid, followed by neutralisation with 0.03 per cent sodium bicarbonate and filtered through a muslin colth, flavoured with essence and colour. The syrup so made is diluted with water

(1 : 4), filled in coloured bottles. These botles are sealed with CO_2 under pressure. The approximate composition of *neera prash* is 11.8 - 14.2 per cent soluble solids, 7.6-8.5 per cent sucrose, 1.6-1.8 per cent reducing sugars and 0.44-0.56 per cent ash (Anon., (Dhanu, India), 1956).

Golden Syrup

It is prepared from *neera, gur* or palm molasses. When prepared from delimed *neera,* the later is inverted with 0.4 per cent citric acid (by weight), brix is brought to 60° by adding water and neutralised using 0.1 per cent sodium bicarbonate (till pH 5.6). When, palm *gur* is used, inversion is done at 25° by brix solution using 0.75 per cent citric acid (on the weight of sugar) and neutralised with 0.35 per cent sodium bicarbonate. When it is prepared from molasses, inversion and decolourisation are affected by using 0.5 per cent citric acid and 1 to 2 per cent activated carbon. The cabon is filtered off after 12-18 hours and neutralisation is done at 50° brix solution using 0.25 per cent sodium bicarbonate. In all cases, the syrup is then filtered at 104°C to separate the sediments. It is then further heated to 109 to 110°C. The product thus obtained is called *golden syrup.* Its composition is given in Table 13.5.

Table 13.5: Composition of Golden Syrup Obtained from Various Sources

Source	Brix per cent	Pol per cent	Surose per cent	Purity co-eff.	Reducing sugars per cent	Ash per cent
Clarified neera	77.7	39.55	44.95	50.9	19.57	1.00
Palm *gur*	80.8	22.4	36.8	27.7	40.2	0.78
Molasses	78.0	52.5	54.4	67.3	12.6	0.65

Source: Anon., (Dhanu, Maharashtra, India), 1956, p. 39.

Refining Sugar from Palm *Gur*

Palm gur is further refined to sugar. In the factory, raw *gur* is heaped on platforms for about two months to drain out most of the molasses. It is then dissovled in water and refined in the usual manner so as to crystallize sugar. The molasses obtained during crystallization is used for producing *arrack.*

Composition of Palm *Gur* and Palm Sugar

Composition of *gur* from saps of various palms and sugarcane juice is given in Table 13.6. *Gur* from various palms has more sucrose, more minerals and relatively less of reducing sugars than the *gur* preparred from sugarcane. Composition of palm sugar and palm *gur* is given in Table 13.7. Palm sugar has relatively more carbohydrates and iron as compared to palm *gur*. Nutritionally, the palm *gur* appears to be superior to palm sugar as it contains relatively more minerals and vitamins. In Indonesia, palm sugar from *Arenga pinnata* contains 62 to 97 per cent sucrose and > 1.5 per cent each of glucose and fructose (Imamkhasani *et al.,* 1989).

Table 13.6: Composition of Palm and Cane *Gur*

Characteristic	Palm *gur* from				Cane *gur*
	Palmyra	Date	Coconut	Sago	
Moisture %	8.61	9.16	10.32	9.16	6.94
Sucrose %	76.86	72.01	71.89	84.31	69.43
Reducing sugars %	1.66	1.48	3.70	0.53	21.18
Fat %	0.19	0.26	0.15	0.11	0.05
Protein %	1.04	1.46	0.96	2.28	0.25
Mineral %	3.15	2.60	5.04	3.66	1.00
Ca as CaO %	0.861	0.363	1.638	1.352	0.400
P as P_2O_5%	0.052	0.52	0.062	1.372	0.045

Source: *Indian Sugar Manual*, 1956, pp. 258; Analysed by Bharatiya Tad *Gur* Shilpa Bhavan Laboratory, Dhanu, Maharashtra, India.

The data indicate that the palm *gur* is more nutritive as compared to the *gur* obtained from sugarcane.

Gur Standard (as per I.S.I.)

Specifications/requirements of *gur* obtained from four palm varieties, namely, palmyrah palm, date palm, sago palm and coconut palm are given in Table 13.8.

Table 13.7: Nutritional Composition of Palm Sugar and Palm *Gur*

Characteristic	Palm Sugar	Palm *gur*
Moisture %	Nil	9.19
Protein %	0.24	0.35
Fat % (ether extraction)	0.37	0.17
Minerals %	0.50	0.74
Carbohydrates %	98.89	90.60
Calcium %	0.08	0.06
Phosphorus %	0.06	0.06
Iron (mg/g)	30.00	2.5
Nicotinic acid (mg/100 g)	4.02	5.24
Vitamin B1 mg/100 g	Nil	24.0
Riboflavin mg/100 g	229.0	432.0
Vitamin C mg/100 g	—	11.0
Calorific value/100 g	398.0	365.0

Souce: *Indian Sugar Manual*, 1956, pp. 257; Analysed by Nutrition Research Laboratory, (ICMR), Coonoor, Tamil Nadu.

Table 13.8: Requirement for Palm Jaggery (Gur)

Sl. No.	Characteristic	Requirement
1.	Pol % by mass (Minimum)	60.0
2.	Moisture % by mass (Maximum)	8.5
3.	Total ash (on dry basis), percentage by mass (Maximum)	40
4.	Water soluble matter (on dry weight basis), percentage by mass (Maximum)	1.5
5.	Acid insoluble matter (on dry basis), percentage by mass (Maximum)	0.5

NB: Total sugars (expressed as invert sugars) on dry weight basis shall not be less than 90 per cent.

Source: Anonymous (1994). *Palm Jaggery (Gur) Specification IS: 13952,* Bureau of Indian Standard, New Delhi, India.

Economics and Marketing

Price of palm *gur* is determined by the quality, especially the flavour. Since the availability of palm *gur* is season bound and the demand is usually more than its production, its price is comparatively much higher than *gur* made from other sources (like those from sugarcane). For example, in West Bengal, the current market price of one kg *Patali gur* (wild date palm *gur*, also known as *Labat* or *Nalen gur*) is Rs. 14-20 per kg depending on the quality, whereas the price of sugarcane *gur* is only

Rs. 8 per kg. On an average, a wild date palm tree can yield 20-25 kg *gur* in a season (during 4 tapping months). Thus, from a single palm a revenue of Rs. 250-400 can be obtained in one season. Misra (personal communication) gave an estimate which suggests that plam *gur* can bring enormous revenue (Table 13.9) provided it is systematically manged. Moreover, a large number of workers (tappers, their family members, and others) can be usefully engaged during tapping period. The picture is similar for other palms also. If all the palms of our country are tapped in a systematic manner, a fair amount of money can be earned/ realised that will support a large number of families in an independent endeavour. Thus, the useful palms should be exploited fully to provide alternative sources of income in rural areas.

Table 13.9: Cost of Production of Wild Date Palm *Gur* and Profit Under West Bengal Conditions

Item	Per plant (in one season)	Per hectare (in one season)
Yield (kg)	20	6000
Gross income (Rs.)	160	48000
Cost of production (Rs.)[1]	80	18000
Profit (Rs.)[2]	80	30000
Cost of production/kg *gur*	4	3

Source: Misra (personal communication).

1. The cost scales down when calculated per ha as efficiency per man hour increases considerably.
2. The profit increases commensurate with scaling up of area over single plant profit.

Some regions are known for producing quality palm *gur* and the price of the produce is usually higher.

The marketing channel for palm *gur* is as follows:

Tappers/ Producers --Middlemen--> Wholesale dealers ------------> Retailers -------------> Consumers

Data collected from different sources revealed that the middlemen usually earn a profit @ Rs. 2-3 per kg of *gur*, the wholesalers earn a profit @ Rs. 2-3 per kg of *gur* (Table 13.10).

Table 13.10: Approximate Estimate of Economic Returns at Different Levels

Return to	Purchase price (Rs./kg)	Sale price (Rs./kg)
Tapper/Producer	—	8.0
Middleman	8.0	10.0 - 12.0
Wholesaler	10.0 - 12.0	12.0 - 15.0
Retailer	12.0 - 15.0	15.0 -20.0
Consumer	15.0 - 20.0*	—

* Depending upon the quality of the produce and season price may be even higher.

Thus, like several other agricultural products, the middlemen take the lion's share of profit, leaving the grower-cum-producer to save for minimum returns doled out to them.

Future Prospects

The source and the technology to produce *gur* from palm make it obvious that chances of its transformation into a massive, well promoted, commercial product is remote unless other sources fail to provide the global needs of sweetener. It serves a much better purpose as a small scale rural industrial product both for the society and the consumer. However, there is need to improve the system of tapping so that the source is not damaged due to excessive or under-tapping. There is further need for improvement in the processing technology through greater mechanisation, improved efficiency, reduced chances of contamination and improve shelf life without sacrificing the 'exotic flavour' that makes it a favourite of so many. Further, the product needs improved market promotion for it to become not only cost effective but also economically lucrative to those who are involved in its production and marketing.

There is an immediate need to work out the genetic and breeding aspects that may provide varieties with higher and sustained yield potential which, in turn, will serve to stabilise the industry without requiring an expansion of land area for cultivation - a resource which is very much in short supply today. Alternatively, these palms can be used as wind breakers as to control soil erosion thus serving a dual purpose. The hardy palms can grow on waste lands as they require hardly any manuring or irrigation. They can easily be grown on bunds of fields, banks of rivers and canals and road sides. For many, the palms are a boon and in fact, date palm is a dominant tree of the desert oases. However, better agro-techniques are needed for higher yields. Similarly, varieties need to be developed that may grow and provide good yield in eroded and low productive soils and other stress conditions where other crops are difficult to grow.

B. GUR FROM SWEET SORGHUM

Sweet sorghum is a short duration crop and contains appreciable amount of sucrose in the stalks. The first effort to manufacture sugar and syrup from sorghum was made in the United States where sorghum was first grown in 1853 from *Chinese Sorgo,* the seeds of which were obtained from France. In 1857, about 16 varieties of African sorghum were introduced into America and these, as a class, were called *Imphee.* Later on, sweet sorghum varieties in the USA were categorised into four groups, e.g., *Amber, Organe, Sumac* and *Gooseneck.* At present sweet sorghum is grown as an alternative sugar crop.in U.S.A., Russia, Germany, Italy, etc., on a small scale.

In India, no classification of saccharine *jowars* exists and among the exotic forms only *Amber* and *Collier* have attracted attention. At the Nimbkar Agricultural Research Institute (NARI) at Phalton (Maharashtra), concerted efforts have been made to make *gur* and sugar from sweet sorghum. Hybrid sweet sorghum variety *Madhura* has been identified with more than 11 per cent sucrose and the ratio of sucrose to reducing sugars is higher than six (Anon., 1993).

Sugar Content in Relation to Growth and Development

Experiments were conducted at Lyallpur (now in Pakistan) and Gurdaspur (Punjab) to elucidate total yield of *gur,* raw sugar and sugar per unit area, quality of the *gur* from sweet sorghum, stage at which the crop should be harvested to yield maximum sugar and the total weight of the crop for fodder.

The high glucose sucrose ratio in juice acts against the production of good crystalline *gur.* The juice also contains substances which often produce insipid *gur*. Further, the yield of *gur* is nearly 1/3 rd of that realised from an average sugarcane crop.

Sugar content in the stripped stalks is maximum when seed contents begin to dry. Thereafter, until the seeds are dead ripe, the total sugar in the crop decreases but the decrease is mainly at the expense of reducing sugars. Because of this glucose ratio seems to decrease as ripening proceeds (Table 13.11).

Table 13.11: Changes in Glucose and Sucrose in Juice in Relation to Growth and Development of Sweet Sorghum

Stage (date of analyses)	Glucose ratio	Sucrose (g/100 cc juice)	Glucose (g/100 cc juice)
A few flowering heads (September 30, 1912)	52.7	2.6	2.90
Almost all heads in flowers (October, 7, 1912)	38.6	4.5	2.82
Seeds in thick milk stage (October, 16, 1912)	24.5	8.6	2.80
Seeds in cheesy stage (October, 21, 1912)	18.9	10.3	2.40
Seed contents nearly dry (October, 28, 1912)	18.6	10.7	2.45
Seeds dry, some quite hard (November, 6, 1912)	15.9	10.8	2.04
Seeds dry, some quite hard (November, 9, 1912)	14.1	10.9	1.92
Seeds dry, many hard (November, 13, 1912)	13.2	10.0	1.53

(Date of planting July, 26, 1912; Seed rate 12 kg/ha)
Source: Annett, 1914.

At Gurdaspur, as the plant ripened, the amount of juice expressed from it decreased. The juice yield was 2177 l/ha, with 13.3 per cent *gur*. As the *gur* contained 57.9 per cent sucrose and 13.42 per cent reducing sugars, it was of poor quality.

In Maharashtra, these days, the sweet sorghum is grown both as Summer Crop and Monsoon Crop. The summer crop is sown in February/March and ripens in May/June when temperatures remain high (> 38°C). The monsoon crop is sown in July and ripens in October. The crop grown during summer generally has low sucrose in the juice (Table 13.12) and, therefore, it is unsuitable for *gur* making (Ghanekar *et al.*, 1978-79).

Table 13.12: Effect of Growing Season and Stage of Harvest on Juice Purity (Sucrose as % Total Solids in Sweet Sorghum)

Stage of harvest	Season	
	Monsoon, 1977	Summer, 1978
Grain milk stage	76.07	49.30
Grain ripe stage	76.18	60.74
CD at 5%	NS	10.23

Source: Ghanekar *et al.*, 1978-79.

In sweet sorghum cultivars HES 4 and HES 13, sucrose content (brix) increased with the age from flowering to maturity. The 4th internode from top had higher sugar content than the top or bottom internodes. In this study, sucrose content (in terms of brix value) was not influenced by the levels of nitrogen and population density. The juice yield was from 9478 to 1,11,176 l/ha^{-1} (Galani *et al.*, 1991).

In Brazil, after flowering as the maturity of plants increases; the percentage of sap and reducing sugars decrease. Total soluble sugars, total sugars, sucrose content and purity increased up to 33 days after flowering and then remained fairly constant with later harvesting dates (Zanini, 1990).

At Stuttgart, in Germany, maximum yield of 22.7 tonnes of dry matter and 6.5 tonnes of free fermentable sugars were obtained at the highest nitrogen rate with adequate water supply. In this study stalk yield increased with increasing rates of nitrogen. Stalk yield was higher in *Rio* as compared to *Dale* (Leible and Kahnt, 1991).

In Texas (USA), in a comparative physiological study of sweet sorghum (*Atlas* × *Rio*) and grain sorghum (*AT* × *623* × *RT* × *5388*), revealed that relatively greater potentials for CO_2 assimilation and ^{14}C-assimilate accumulation in ripe stem tissues were associated with higher levels of stem non-structural carbohydrates in sweet sorghum as compared to grain sorghum. After physiological maturity, the concentration of non-structural carbohydrates in the lower and upper part of the stalk of sweet sorghum was 1.4 and 2.7 times higher than that of grain sorghum (Vietor and Miller, 1990).

In India, Ghatode *et al.*, (1991a) studied the influence of cultural practices and observed that delay in sowing (from June last week onwards) decreased stalk diameter. The total soluble solids, reducing sugars, non-reducing sugar contents decreased with delay in harvesting up to maturity stage, while juice acidity increased.

A common belief prevailed amongst farmers that cultivars possessing plain white mid rib were less sweet than those having more or less inconspicuous mid rib. A concerted effort to unravel this issue did not reveal any obvious connection (Annett, 1914).

Sucrose Content and the Stalk Yield

Stalk yield and sucrose per cent in the juice extracted from the stalks of various exotic sweet sorghum varieties under Maharashtra conditions are given in Table 13.13. The data in the table indicate that the sweet sorghum variety *Rio* performed relatively better than the other exotic varieties.

Table 13.13: Stalk Yield and Sucrose per cent in the Juice of Sweet Sorghum cultivars

Variety	Yield of stripped stalks ($t\ ha^{-1}$)	Sucrose % juice
Rio	32.43	10.61
Brawley	29.43	10.57
Tracy	21.89	9.37
Collier	20.00	10.41
Rex	19.44	9.98

Source: Annl. Rept. Nimbkar Agricultural Research Institute, Phaltan, Maharashtra (1969-70), pp. 15 - 16.

In Korea, yields of sweet sorghum vary from 43.3 to 150 $t\ ha^{-1}$ with a sucrose content of 6 to 14 per cent (Park and Lee, 1991). In Quebec (Canada), *S. bicolor* var. *saccharatum, Roma* and *Rio* produced on an average 41 t fresh stalks ha^{-1} with sugar content in the juice being 16° Brix (Chamberland, 1976).

Not only the plant crop but the ratoon crop of sweet sorghum also gives appreciable yield of sugar. In U.S.A., fresh weight of ratoon crop is 36 per cent higher but the sugar concentration is 15 and 9 per cent lower in the first and second ratoon than the corresponding plant crop (Duncan and Gardner, 1984).

Cultural Practices and Sweet Sorghum Juice vis-a-vis *Sugarcane Juice*

Cultural practices for growing sweet sorghum for *gur* and the yield levels obtained are given in Table 13.14.

Table 13.14: Cultural Practices for Growing Sweet Sorghum *Madhura* Variety for *Gur* and Syrup Production

Date of planting	April - May	Mid November - Mid January
Crop duration (days)	110 - 120	120 - 130
Seed rate (kg/ha)	7.5 to 10.0	
Plant density	1.5 to 2.0 lakh plants/ha; spacing: row-to-row: 45 cm, plant-to-plant: 15-20 cm	
Fertiliser dose (kg/ha)	N:P:K (100: 50:50) at planting; N (50) one month after planting	
Number of irrigations	4 - 6	8 - 10
Irrigation interval (days)	8 - 10	13 - 15
Yield levels		
(a) Stripped stalks	60-80 t / ha	
(b) Grain	2-4 t / ha	

Source: Rajvanshi *et al.*, 1993.

A comparison of the juice characteristics of sweet sorghum (variety *Madhura*) and sugarcane (variety Co 740, normally used for *gur* making) (Table 13.15) indicates that brix and sucrose in juice are of the same order.

Effect of Post-harvest Storage on Juice Quality

Studies on the effect of post-harvest storage of sweet sorghum (cultivar : SSV - 108) stalks on juice quality and syrup recovery indicated that juice extraction decreased with storage time. The highest juice extraction (31.1%) was realise from fresh stalks at the ripe stage, and the lowest (0.0%) from post-ripe sorghum after 120 hours of storage. The highest total soluble solids and reducing sugars were recovered at the ripe stage after storage for 144 hours. The highest syrup recovery was recorded at the ripe stage after 48 or 72 hours of storage (Ghatode *et al.*, 1991b). Sweet sorghum varieties differed with respect to post-harvest deterioration. In a study, conducted in Maharashtra, brix at harvest was highest (17.0) in HES 13 and lowest (9.7) in HES 3; weight loss (average for all the storage periods) was lowest (5.12%) in HES 13 and highest 8.98 per cent in HES 3. Over a four-day period, after harvest, weight loss averaged 15.73 per cent, average juice yield decreased from 41.7 to 37.6 per cent and average brix value increased from 14.4 to 18.0 (Choudhary, 1991).

Table 13.15: Characteristics of Sweet Sorghum and Sugarcane Juice

Characteristic	Sugarcane (Co 740)	Sweet sorghum (*Madhura*)
pH	5.8 - 6.4	4.7 - 5.5
Brix (□)	16 - 22	12 - 21
Sucrose (% w/w)	13-18	10-15
Reducing sugars (%, w/w)	0.2 - 1.0	0.75 - 3.0
Starch (%, w/w)	0.001 - 0.05	0.02 - 0.22
Aconitic acid (%, w/w)	0.01 - 0.03	0.08 - 0.56
Minerals (%, w/w)	0.4 - 0.7	0.4 - 0.5

Source: Rajvanshi *et al.*, 1993.

Manufacture of Gur from Sweet Sorghum

Juice from the stalks of sweet sorghum contains 9-11 per cent sugars (Anon., 1969-70; El Bassam *et al.*, 1987) and amongst sugars there are 53 per cent sucrose, 28 per cent glucose and 19 per cent fructose (El Bassam *et al*, 1987).

Efforts to prepare *gur* from sweet sorghum in India were first made by H.E. Annett who succeeded partially in making *gur* from sweet sorghum on October 23, 1912. It was of low consistency, watery and of little value. On November 5, 1912, semi-solid *gur* was obtained. The taste of the product was so aweful that even cattle did not eat it. On November 9, 1912, some success was achieved in making better quality *gur* from sweet sorghum (Annett, 1914).

Juice from stalks of three-month-old crop (var. *Rio*) was clarified using *deola (Hibiscus abelmoschus)* at the rate of 0.8 kg per tonne of stripped stalks. The clarified juice is then concentrated till it reaches the striking point (110-112°) and is allowed to cool in wooden trays to form lumps of *gur* (Ghosh, 1972).

The Nimbkar Agricultural Research Institute, Phaltan (Maharashtra), has developed an appropriate technology for mass scale manufacture of *gur* from sweet sorghum. The stalks of sweet sorghum are harvested when grain is in a soft dough stage, and crushed in an ordinary sugarcane cusher after the removal of leaves and sheaths. The pH of extracted juice is about 5.5 and it is neutralized by adding an adequate quantity of $Ca(OH)_2$. The neutral juice is placed into a boiling pan and mucilage of okra (*Abelmoschus esculentus*) is added to juice. Thereafter, it is heated to a temperature of 55-60°C. Convection currents, developing in the juice, bring the starch particles to the surface where they get trapped in the mucilage. After about 30-40 min, a thick layer of greenish sucm is formed over the surface of the juice. The scum is removed with a perforated laddle. Thereafter, heat is increased to boil the juice up to the striking point. At this stage, the thickened juice is poured into a congealing pan. The sugary mass is allowed to cool under constant stirring and when crystallization begins, it is shoveled into moulds, having perforations for allowing the non-crystallizable reducing sugars to ooze out. The moulds are allowed to stand for 12 hr for the formation of solid *gur* (Anon., 1973-74).

In Maharashtra, Bapat *et al.* (1987) evaluated the performance of sweet sorghum cultivars with regards the production of quality syrup and *gur*. Variety SSV 252 had the highest millable stalk height (344 cm) and green stalk yield (57.6 t ha^{-1}) followed by SSV 74 and SSV 7073 (52.2 and 51.7 t ha^{-1}). The variety SSV 7073 possessed the highest brix of 22.2 and pol per cent juice of 15.9. *Gur* prepared from juice of SSV 84 and SSV 108 contained 64.4 and 63.7 per cent non-reducing sugars and 12.5 and 9.3 per cent reducing sugars, respectively.

In a recent press release from the Nimbkar Agricultural Research Institute, Phaltan, it has been stated that out of 3 varieties of sweet sorghum evaluated for *gur Madhura* has performed the best (Anon., 1993). A pilot plant of 100 kg capacity has been set up wherein a bio-gasifier (50-90 kilo watts) is run on loose biomass to make *gur*. In this study, 48 per cent juice extraction and 8 per cent recovery of *gur* have been recorded (Fig. 13.11).

Maximum scum removal (13-15% (w/w) of juice) takes place when the boiling is achieved within 10-15 minutes. No extra flocculent or clarificant is required. Best *gur* is made when the striking point is between 114-116°C. Setting of *gur* is relatively better when the cooling time of the heated mass is between 7-10 minutes. Because of the lower sucrose: reducing sugars ratio in sweet sorghum, more vigorous working of the concentrated mass is required during cooling period for better colour and quality of *gur*. It has been observed that *gur* produced in aluminium utencils made (*Kadhai*) is of better colour than the one prepared in utencil of mild steel. Relatively better syrup with good colour and flowability has been prepared when the final brix reached 70-75° (corresponding to a syrup temperature of ~105°C (Rajvanshi *et al.*, 1993).

Comparison of Gur made from Juice of Sweet Sorghum and Sugarcane

A comparative study of *gur* prepared from the juice extracted from sweet sorghum and sugarcane was made by Ghosh (1972) (Table 13.16). Rajvanshi *et al.* (1993) also compared the major constituents of *gur* from sweet sorghum (variety *Madhura*) and sugarcane (variety Co 740) (Table 13.17).

The *gur* prepared from sweet sorghum contains relatively less sucrose, more reducuing sugars, more sulphate ash and more colour than the one prepared from sugarcane. Organoleptic tests have also showed that the *gur* from sweet sorghum is *at par* with the one prepared from sugarcane.

There are some other plus points in making *gur* from sweet sorghum. These are: (a) in a year, *gur* from *Madhura* variety could be made for 160 days from April to June and again from August to November, while sugarcane based *gur* units run from October to March, and (b) sweet sorghum is a relatively short duration crop requiring less of water (Anon., 1993; Rajvanshi *et al.*, 1993).

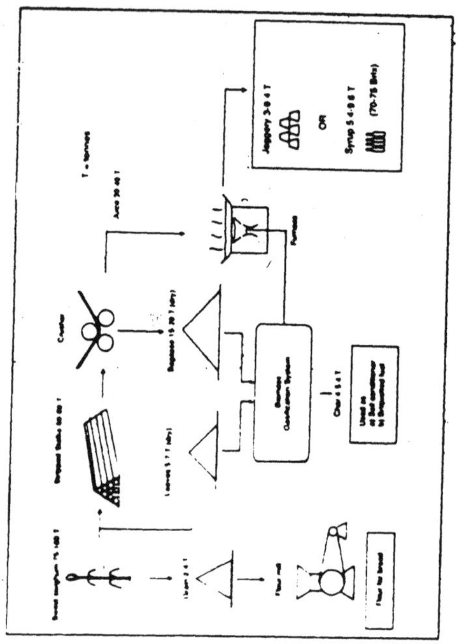

Fig. 13.11: Schematic Diagram for the Manufacture of *Jaggery*, Syrup and Grain from Sweet Sorghum Hybrid *Madhura*.
Source: Rajvanshi *et al.*, 1993)

Table 13.16: Comparison of *Gur* from Sweet Sorghum and Sugarcane

Constituent	Gur prepared from	
	Sweet sorghum	Sugarcane*
Moisture (%)	6.52	6.24
Hardness (Unit)	3.10	3.30
Sucrose (%)	78.10	84.20
Reducing sugars (%)	8.8	7.5
Colour in solution	378**	305**
Colour of solid *gur*	Plate 11,45 J 7***	Plate 11,45 F***
Crystal (%)	58.1	64.2
Sulphate ash (%)	7.6	4.3
pH	5.5	5.6

* Analysis of representative *gur* samples.
** Colour is determined in N/2 *gur* sloution using 540 nm filter.
*** From *A Dictionary of Colour* (Maerz and Rea Paul, 1950).
Source: Ghosh (1972).

Table 13.17: Major Constituents of *Gur* from Sweet Sorghum and Sugarcane (wet basis)

Constituent	Sweet sorghum (*Madhura*)	Sugarcane (Co 740)
Moisture (%)	7 - 13	9 - 11
Sucrose (%, w/w)	53 - 69	70 - 78
Reducing sugars (%, w/w)	8 - 23	7 0 10

Source: Rajvanshi *et al.*, 1993.

Constraints in Manufacturing Gur from Sweet Sorghum

The main problem in making *gur* from sweet sorghum is the presence of starch in juice (0.33 to 1.4%); this interferes with the process of solidification and crystallization of the sugar. Following methods have been tried for the removal of starch at the Nimbkar Agricultural Research Institute, Phaltan, Maharashtra:

(a) Enzymic hydrolysis of starch by using enzyme amyloglucosidase,
(b) Sedimentation of starch and other colloidal particles by flocculating agents, namely, filter floc or Magnesia preparation, and

(c) Flotation of starch and other colloids and removing them as scum.

All these were successful on a laboratory scale but (c) appeared to be the best method for mass scale manufacture of *gur* (Anon., 1973-74; Karve and Ghanekar, 1974). Rajvanshi *et al.* (1993), however, observed that starch and aconitic acid normally present in the sweet sorghum juice of *Madhura* variety do not affect *gur* making.

Syrup Preparation

For preparing syrup, juice is boiled in a shallow iron pan using dried bagasse as a fuel. For clarifying juice, *sukhlai* aqueous extract is added before boiling. The scum is carefully skimmed off and a fine golden yellow clear liquid is finally obtained. It takes approximately two hours to boil the liquid down to a proper density.

Syrup obtained from clarified sweet sorghum juice (using 6 g activated charcoal/100 ml of juice) is amber coloured and sweet. However, unclarified syrup remains dark brown and salty (Krishnaveni *et al.*, 1990).

From the juice of sweet sorghum (variety *Madhura*) excellent syrup with good colour and flowability has been prepared when the final brix reached 70-75° (corresponding to a syrup temperature of ~105°C (Rajvanshi *et al.*, 1993).

Shelf Life of Sweet Sorghum Gur and Syrup

Shelf life of sweet sorghum *gur* (variety *Madhura*) is similar to the one prepared from sugarcane. However, during storage of syrup, partial crystallization and fungal invasion occurs. Storage of the stalks before juice extraction may prolong the shelf life of the syrup (Rajvanshi *et al.*, 1993).

Improving Efficiency of the Furnace Used in Gur Manufacture

Existing furnaces are normally very inefficient (10-15% overall efficiency). One of the major improvements in *gur*-making can be effected by improving the furnace efficiency. At the Nimbkar

Agricultural Research Institute, Phaltan, Maharashtra, a multi-fuel biogasifier has been developed which can run on sweet sorghum bagasse, sugarcane leaves, wheat husk, dry grasses, etc., (Rajvanshi and Jorapur, 1992; Rajvanshi *et al.*, 1993). A pilot gasification plant of 100 kW (thermal) capacity produces about 15-20 kg of *gur* per batch. The advantages of this gasifier over the conventional furnace are: (a) it is fully automated system, (b) produces a bluish white, smokeless flame having temperatures higher than 1250°C, (c) overall efficiency (biomass to useful heat) of the system is 20-27 per cent which is much higher than the conventional furnaces, (d) this system produces char as by-product (20-25% (w/w) of fuel). After mixing with suitable binder (15% (w/w) cow dung), this produces excellent smokeless fuel briquettes for domestic consumption. Besides, it also acts as a soil conditioner (Rajvanshi *et al.*, 1993).

Economic Analysis

A detailed economic analysis for *gur* and syrup production from sweet sorghum variety *Madhura* has been given by Rajvanshi *et al.* (1993). They have computed the cost of sweet sorghum *gur* production to be Rs. 6.80/kg (assuming a sweet sorghum cost of Rs. 200 per tonne and 8 per cent recovery). The corresponding syrup cost comes to Rs. 4.56/kg assuming a recovery of 12 per cent on stalk basis.

Factory Scale Trials of Sugar Production from Sweet Sorghum

Factory scale trials of sugar production from sweet sorghum in Mexico indicated that there was no processing problem with freshly harvested sweet sorghum stalks but clarification becomes difficult when there was a gap of two days between harvesting and milling. Low jucie purity was attributed to the harvesting of immature stalks (Coeto and Smith, 1974).

Chatterjee *et al.* (1975) have described a method for producing white sugar from sweet sorghum *gur*. The *gur* is melted to 30° Bx semi-syrup, which is treated with lime-phosphoric acid and a flocculent is added to precipitate the starch. It is then concentrated to 65° Bx and lime is added to

attain a pH of 8.0 to 8.2. Calcium chloride is added to precipitae aconitic acid. Now 4 : 1 mixture of sorghum syrup : cane syrup is boiled to produce raw sugar at pol of 90.8. This raw sugar is remelted and reboiled to give white sugar of pol 99.8.

REFERENCES

Annett, H.E. (1914). The use of sweet jowar (*Sorghum* sp.) as a source of commercial sugar or as a fodder and the variation in composition of the crop during growth. *Bulletin No. 41,* Agricultural Research Institute, Pusa, Superintendent, Government Printing, Calcutta, India, pp. 9.

Anonymous, (1969-70). *Annl. Rept.,* Nimbkar Agricultural Research Institute, Phaltan, Maharashtra, India.

Anonymous, (1973-74)., *Annl. Rept.* Nimbkar Agricultural Research Institute, Phaltan, Maharashtra, India.

Anonymous, (1993). Sweet sorghum for the sweet tooth. *Times of India* (Lucknow), dated January, 18, 1993, p. 7.

Anonymous, (1940). *Annl. Rept., 1939, Coconut Research Scheme,* pp. 5.

Anonymous, (1978). World production of non-centrifugal sugar. *Encl. Britannica* (Macropaedia) 5: 938.

Anonymous, (1978). World production of non-centrifugal sugar. *Encl. Britannica* (Micropaedia) VIII: 699.

Anonymous, (1994). Palm Jaggery (*Gur*) Specification IS : 13952 : Bureau of Indian Standard, New Delhi, India.

Anonymous, (Anakapalle, India) (1989). *75 Years of Agricultural Research (1913-1988) at Anakapalle,* Regional Research Station, Anakapalle, Andhra Pradesh, A.P.A.U., Hyderabad, A.P., pp. 400.

Anonymous, (Dhanu, India) (1956). *Laboratory Investigational Work on Palm Juices and Gurs (1953 to 1955),* Bulletin No. 1, Bharatiya Tad Gud Shilpa Bhava, Dhanu, All India KVI Board, Mumbai, India, pp. 44.

Bailey, L.H. (1930). *Hortus Second* (Compiled by L.H. Bailey and E.Z. Bailey), The Macmillan Co., New York.

Bapat, D.R., H.D. Jadhav, S.L. Gaur and C.B. Salunkhe (1987). Sweet sorghum cultivars for producion of quality syrup and jaggery in Maharashtra, in *Technology and Applications of Alternative Uses of Sorghum, Proc. National Seminar, Parbhani, Maharashtra, Feb., 2-3, 1987* (eds. Ingle, V.M., D.N. Kulkarni and S.S. Thorat), M.P.A.U., Rahuri, Maharashtra, pp. 203.

Barber, C.A. (1918). Progress of the Sugarcane Industry in India during the years 1916 and 1917. *Bulletin No. 83,* Agricultural Researh Institute, Pusa, Superintendent, Government Printing, Calcutta, India, pp. 45.

Blatter, E. (1926). *The Palms of British India and Ceylon,* Oxford Univ. Press, London and New York.

Bowles, R. (1990). Tropical tree of plenty. *Farmers Weekly* , 111 (8) : 8.

Chamberland, E. (1976). [Trials of sweet sorghum in Qubec]. *Naturaliste Canadien,* 103(5) : 543-551.

Chatterjee, A.C., S.N. Joshi and K.B. Nadgauda (1975). White sugar production from sweet sorghum. *Proc. 5th Jt. Conv. S.T.A.I., D.S.T.A., S.I.S.S.T.A.*: M-73-M-79.

Choudhary, S.D. (1991). Effect of time lag after harvesting for cane crushing on juice and brix in genotypes of high energy sorghum. *Bharatiya Sugar*, 16(4) : 39-40.

Coeto, J.A. and B.A. Smith (1974). Factory-scale tests of sugar production from sweet sorghums in Mexico. *Proc. International Soc. Sugar Cane Technol.* 15: 1356-1362.

Corner, E.J.H. (1966). *The Natural History of Palms*, Berkeley and Los Angeles, University of California Press.

Davis, T.A. (1977). The wild Date-palm. *Indian Farming*, 26(11) : 90-92.

Duncan, R.R. and W.A. Gardner (1984). The influence of ratoon cropping on sweet sorghum yield, sugar production and insect damage. *Can. J. Pl. Sci.*, 64(2) : 261-273.

El Bassam, N., M. Dambroth and G. Ruhl (1987). [Sweet sorghum, a new raw material for sugar industry]. *Landbauforschung Volkenrode*, 37(4) : 201-206.

Galani, N.N., M.H. Lomte and S.D. Choudhari (1991). Juice yield and brix as affected by genotype, plant density and nitrogen levels in high energy sorghums. *Bharatiya Sugar*, 16(4) : 23-24.

Ghanekar, A.R., S.H. Kshirsagar and A.D. Karve (1978-79). *Annl. Rept.* Nimbkar Agricultural Research Institute, Phaltan, Distt. Satara, Maharashtra 12 : 43-44.

Ghatode, K.S., V.B. Kalmegh, B. Sagare and N.R. Pavitrakar (1991a). Effect of environment on quality of sorgho and its juice in relation to phenological stages. *Annl. Pl. Physiol.*, 5(1) : 52-57.

Ghatode, K.S., V.B. Kalmegh, A.B. Kadu, P.N. Nimbkar and N.B. Gore (1991b). Effect of post-harvest conditioning of sorgho stalks on juice quality and syrup recovery at various phenological stages. *Annl. Pl. Physiol.* 5(2) : 238-241.

Ghosh, A.K. (1972). *Gur* from sweet sorghum. *Indian Sugar*, 22(1) : 1-2.

Griffith, W. (1850). *Palms of British East India*. Govt. of Bengal, Calcutta.

Hawakes, A.D. (1950-52). *The Major Kinds of Palms* (Botanical papers 1-8, Fairchild Tropical Garden, Coral Gables, Florida, USA.

Imamkhasani, S., J. Kantasubrata and S. Sumartini (1989). Analysis of mono- and disaccharides in production of palm sugars by a high performance liquid chromatographic method. *J. Chromatographic Sci.*, 27(11) : 676-679.

Karve, A.D. and A.R. Ghanekar (1974). Clarification of sweet sorghum juice for jaggery making. *Proc. Annl. Conv. D.S.T.A.*, 26(1) : M 51-M53.

Krishnaveni, S.,T. Balasubramaniam and S. Sadasivam (1990). Potentiality of sweet sorghum (*Sorghum bicolor*, Poaceae) for syrup preparation and alcohol production in India. *Econ. Bot.*, 44 (3) : 355-359.

Kulkarni, D.G. (1971). *Tropical Sugar*, Dhyan Jyot Prakashan, Poona, India.

Leible, L. and G. Kahnt (1991). [Investigations into the effect of location, sowing rate, N application, cultivar and harvesting date on yield and composition of sweet sorghum. *J. Agronomy and Crop Sci.*, 166(1) : 8-18 (in German).

Loomis, H.F. (1949). *The Nipa Palm of the Orient* (Reprinted from *National Horticultural Magazine*, January, 1949).

Maerz, A. and M. Rea Paul (1950). *A Dictionary of Colour*, McGraw-Hill Book Co., Inc., New York, Toronto and London, pp. 208.

Marsden, H. and F.L. Skilton (1931). Yields of toddy from coconut plants. *Malay Agric. J.*, 19 : 287-290.

Mc Currach, J.C. (1960). *Palms of the World*, Happer and Brothers, New York.

Menon, K.P.V. and K.M. Pandalai (1960). *The Coconut Palm—A Monograph*, Indian Central Coconut Committee, Ernakulam, India, pp. 293-304.

Mitra, J.N., D. Mitra and S.K. Chaudhri, (1990). *Studies in Botany*, Vol. I, Moulik Library, Calcutta, pp. 1021.

Nathanael, W.R.N. (1956). Toddy yields from coconut palms in Ceylon. *Ceylon Coconut Quart.*, 6 : 8-16.

Omran, H. (1988). [Production of nira and brown sugar from coconut cultivar Mawa]. *Teknologi Koko-Kelpa*, 3(1) : 45-48.

Park, K.B. and M.H. Lee (1991). [Feasibility of utilization of sugar crops as bio-energy resources in Korea]. *Korean J. Crop Sci.*, 36(4) : 300-304. (in Korean Language).

Patel, J.S. (1938). *The Coconut - A Monograph*, Govt. Press, Madras.

Purseglove, J.W. (1975). *Tropical Crops - Monocotyledons*, Vol. I and II (ELBS edn.), Longman Group Ltd., London, pp. 428.

Rajvanshi, A.K. and R.M. Jorapur (1992). *Development of leafy biomass gasification systems. Publication No. NARI-GS-1* (1992), Nimbkar Agricultural Research Institute, Phaltan, Maharashtra, India.

Rajvanshi, A.K. and D K. Tapas, R.M. Jorapur and N. Nimbkar (1993). *Jaggery and syrup from sweet sorghum. Publication No. NARI-GUR* (1993), Nimbkar Agricultural Research Institute, Phaltan, Mahrashtra, India, pp. 10.

Roy, S.C. (1951). *Monograph on the Gur Industry in India*, Indian Institute of Sugar Technology, Kanpur, India, pp. 300 + LXXIV.

Samba, Murthy, A.V.S.S. and N.S. Subrahmanyam (1989). *A Text Book of Economic Botany*, Wiley Eastern Ltd., pp. 269-272.

Vietor, D.M. and F.R. Miller (1990). Assimilation, partitioning, and non-structural carbohydrates in sweet sorghum compared with grain sorghum, *Crop. Sci.*, 30(5) : 1109-1115.

Walawalkar, D.G. (1950). Sulphanilamide as a preservative for sweet juices from palm trees. *Nature*, 165 : 370.

Zanini, J.R. (1990). [Influence of physiological maturation on production and quality of seeds and on industrial yield of sweet sorghum]. *Pesquisa Agropecuaria Brasileira*, 25(6) : 881-888 (in Portuguese).

14

SUGARCANE CRUSHERS

The sugarcane which is being grown in India from times immemorial has been classified under two main groups. The thin hardy canes of north India come under *Saccharum barberi* and *Saccharum sinense* and the thick juicy ones (*Saccharum officinarum)* come under tropical India.

Sugarcane Crushers in the Ancient Period

Literature indicates that China was one of the earliest countries where the cane was grown and sugar was made. In the period of the Northern Sung Dynasty (960-1126), Wang Shao, a native of Shi Ning in Szechuan Province, wrote a book *Tang Shuang Pu*, meaning 'A Treatise on Sugar Frost' is perhaps the first exclusive book on sugarcane and sugar, in which origin of sugarcane varieties, crushing cane to extract juice and the methods to produce crystallised sugar are summarised. Chang (1992) has given extract from this book which describes crushing as follows:

> Sugarcane was peeled, cut into small pieces and milled to extract juice. The mill was made of two round stones, one placed above the other. The upper one, being bullock drawn part was 7 or 8 feet long weighing more than one thousand pounds. The lower stone, the base, had a circumference of more than ten feet. In some other cases a mortar was used. Small pieces (about the size of a coin) of sugarcane were extracted through the stone mills, then pressed in a press made up of bamboos. The juice was collected in lacquered jars.

Contribution from: Dr. N.S.L. Srivastava, Central Institute of Agricultural Engineering, Bhopal, Madhya Pradesh, India

Later on, during the Ming Dynasty, in 1637, Sung Ying Hsing wrote a great treatise on agricultural and industrial technology *Tien Kung Kai Wu* in which sugarcane crushing has been described as follows: The cane juice was extracted in a primitive wooden crusher consisting of two horizontal platforms; at the four corners of which were inserted wooden pillars well buried in the ground to stabilise the crusher. At the centre of the upper platform there were two holes, holding two deeply corrugated rolls of hard wood. The one protruding through the upper platform was connected to a piece of bent wood, driven by an ox. In this way, one of the rolls was in motion, while the other remained stationary. The two toothed rolls worked in perfect unison, so that the sugarcane passing through was properly crushed. The juice coming out was collected into a jar placed under the crusher. The cane residue was crushed again so as to extract more juice. After three crushings the juice was largely extracted (Chang, 1992).

Gazetteer of Oudh, India (1877-78) describes a crude mill which is nothing more than a sharpened end of a small tree, working with a rotary motion inside a partially hollow log. It was unable to express the juice from a finer kind of cane, while the inferior one readily yields to the pressure brought to bear upon it.

In the eastern part of the then United Provinces, stone crushers, in the form of a huge decorated structure were used even in the earlier part of this century. Some of these, though have become extinct, as they were more energy intensive and less efficient (25 to 30 per cent) in crushing may still be seen lying idle in eastern districts of Uttar Pradesh. A two roller vertical wooden crusher used around 1750 A.D. in India is depicted in Fig. 14.1

In the erstwhile Punjab province, in the earlier part of this century, wooden crashers were used. These were popularly known as *belna*. These have almost disappeared during the second decade of this century. Though some of these were seen in some parts of Hoshiarpur and Jalandhar (Barnes, 1918).

Till 1925, a large number of sugarcane crushers were imported from England, USA, France, etc. Later, M/S Kirloskar Brothers, M/S Nahan Foundry, Indian Institute of Sugar Technology (now National Sugar Institute) and several other

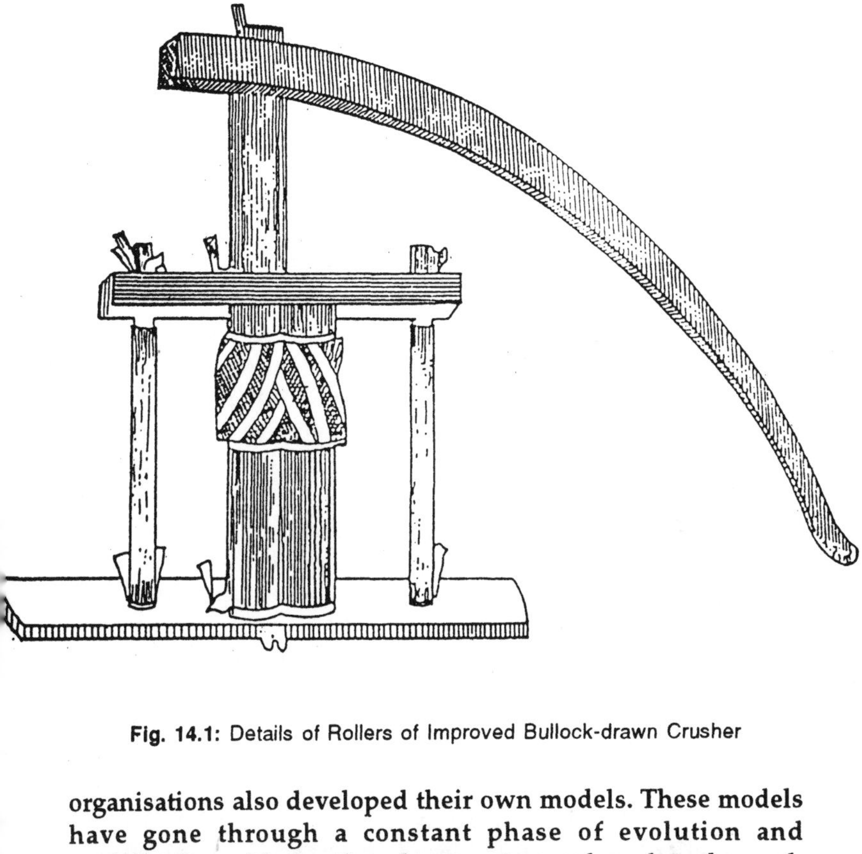

Fig. 14.1: Details of Rollers of Improved Bullock-drawn Crusher

organisations also developed their own models. These models have gone through a constant phase of evolution and modification with time in subsequent years based on the work experience and are now popular in the country.

Bullock drawn sugarcane crushers have become more popular from the beginning of the 20th century. Initially animal drawn crushers were having only two rollers, king and extraction, but later on one more roller (feed roller) was added to the design and three-roller crushers became more common. These designs were worked out empirically on the basis of experience over a long period of time.

It was due to simplicity that the bullock drawn crushers were provided with vertical rollers. The king roller meshes with the other two rollers with the help of spur gears. The king roller is connected with a long wooden beam about 4.5 m in length, with the help of a mounting box. Power is transmitted to the king roller by hitching a pair of animals with the beam with the help of a rope. The cane is fed between king and feed rollers where they are crushed and nodes and internodes are ruptured. The material then passes between the king and extraction rollers and final extraction of juice is obtained by putting more pressure on the cane. This is accomplished by reducing the gap between the rollers.

Power operated sugarcane crushers became popular approximately by the middle of 20th century. They are operated by a small 3.75 to 7.5 kW engine or electric motor. Bigger size power operated sugarcane crushers are also being used.

Extraction of juice depends on the rupture of the juice containing cells which, in turn, is dependent on the pressure applied on the cane between the rollers. Higher the pressure more is the extraction. The capacity of the crushers is dependent on the speed of the rollers, gap between rollers and feeding rate. Higher the speed and gap between rollers more is the crushing rate but with reduced extraction.

In this chapter work related to the performance of small sugarcane crushers has been reviewed. Information available on the performance of sugar mills and *khandsari* units which are relevant in case of small crushers has also been reviewed.

Storage of Sucrose in the Stalk

The sugars synthesised in the leaves are partly utilised in the production of energy and partly in the building up of plant body (van Dillewijn, 1952, Alexander, 1972). The remainder is stored in the stalk in the form of sucrose by a process known as ripening. Translocation takes place through the phloem elements. Sugar is stored in the vacuole as well as in the cytoplasm of the parenchymatous cell (Thom and Maretzki, 1989).

A cane stalk consists of many joints, their age varies according to their formation. Ripening is a joint to joint process. The top joints are less mature than the bottom ones. Mature cane is characterised by a more or less uniform sucrose content along with the stalk with the exception of the top portion of the cane and the below ground stubble. The low sucrose content in the top joints is due to the fact that the joints of this portion are not fully grown, while the lower sucrose content of the stubble is associated with high fibre content. The trend of glucose content is almost contrary to that of sucrose. It is highest in the youngest internode and decreases as the internodes grow older. In a mature cane, glucose content is very low except in the top portion. The cane is considered mature when the brix in juice is about 18 per cent.

Rind Hardness and Fibre Content

The rind consists of several layers of cells mostly sclerenchymatous (van Dillewijn, 1952). Next to the epidermis occurs two rows of thick-walled and lignified cells which give strength to the stalk. Next to these cells there are one or more rows of thin-walled parenchymatous cells which contain chlorophyll. Below lies cortex, whose width and composition vary in different regions of the stalk.

With increase in age, there is more and more deposition of lignin-like compounds in and around fibrovascular tissues, resulting in what might be called a hardening of the fibre up to the time of tasseling (van Dillewijn, 1952; Davidson, 1968; Martin and Cocharan, 1975; Mungomery, 1936; Skinner, 1974). After tasseling, there is a sudden change in the activity of these cell groups from deposition to that of removal of sucrose. Much of the lignin is lost from the fibrovascular tissues causing the bundles to become more flaccid than before. The rind becomes softer and is more easily penetrated because the remaining tissues are mainly cellulosic in nature. In the internodes, the vascular bundles run almost parallel to each other but in the nodes, many of them branch and reanastomose to go to the leaves, buds or root primordia. Because of this the nodes are much thicker and harder than the internodes.

The hardness of the cane stalks, a varietal character, is of considerable importance both in the mill and in the field. Harder canes often pose problems in milling because of sudden breakage of mill rollers. In the field, rind hardness is closely associated with the damage caused by squirrels, rats, jackals, stalk borers, and even by the human beings. Softer varieties are damaged more frequently than the harder ones.

For measuring hardness simple devices have been developed by Puri and Venkatraman (1929). The device consists of a plunger mounted on a stand. The upper end of the plunger is provided with a circular plate to carry weights while the lower end with a piercing bit of 0.7 mm in diameter. The degree of hardness is measured in terms of pounds or kg of weight required on top of the plunger to pierce the rind of a cane. Khanna (1935) developed a similar instrument for field use based on the principle of "pressometer".

Rind hardness shows a wide range of variation caused by both internal and external factors. Puri and Venkatraman (1929) found the rind hardness of *Keli,* the softest tropical cane as 2.5, other tropical canes 4 to 6, indigenous Indian canes 7 to 8, hard rind Coimbatore canes as 8 to 9 and *Saccharum spontaneum* as 10. Non-irrigated canes have higher rind hardness as compared to the irrigated ones.

Rind hardness within one stalk may vary greatly and the data available are rather conflicting as many factors are involved. However, crushing a harder rind would require more force than the softer one.

Composition of Sugarcane and Juice

The trash free cane stalk is composed of approximately 75 per cent water and the remainder is divided between fibre and soluble solids (Table 14.1). The amount of these three components is genetically determined and varietal differences are well documented. The *noble* varieties are rich in water and have low fibre and sugar. The hybrid varieties, now more popular world over, have both high fibre and sugar.

Most of the Indian varieties of sugarcane contain about 84 per cent absolute juice consisting of about 65 per cent water, 15.5 per cent sugar, 3.5 per cent non-sugars and 16 per cent fibre (Ramana Rao *et al.,* 1979). Higher the fibre content in cane higher is the power requirement for the extraction of juice.

Types of Sugarcane Crushers

The present day sugarcane crushers in vogue can be broadly classified into three categories:

(a) Manually operated crusher

(b) Animal operated crushers, and
(c) Power operated crusher.

Table 14.1: Composition of Sugarcane and Solids in the Juice

Component	Composition (% cane)	
	Noble varieties	Indian hybrid varieties
Millable cane		
(a) Water	73 - 76	65
(b) Solids	24 - 27	35
Soluble solids	10 - 16	16
Fibre (dry)	11 - 16	
Juice constituents (% soluble solids)		
(a) Sugars	75 - 92	
Sucrose	70 - 80	
Glucose	2 - 4	
Fructose	2 - 4	
(b) Salts	3 - 4.5	
Inorganic acids	1.5 - 4.5	
Organic acids	1.0 - 3.0	
(c) Organic acids		
Carboxylic acids	1.1 - 3.0	
Amino acids	0.5 - 2.5	
(d) Other organic non-sugars		
Proteins	0.5 - 0.6	
Starch	0.001 - 0.050	
Gums	0.3 - 0.6	
Waxes, fats, Phosphatides	0.05 - 0.15	
(e) Others	3.0 - 5.0	

Development in Sugarcane Crushers

The ancient crushers are followed by two roller crushers made of wood or iron. The canes were normally fed repeatedly through these rollers to extract the juice. From some of the experiments conducted in earlier part of this century (Table 14.2) one can infer that the old wooden mill, the *belna*, is less efficient in extracting juice than the 3-roller Nahan Mill at Gurdaspur (Barnes, 1918).

In wooden *belna* around 20 to 33 per cent of the juice is left unextracted representing a loss of 20 per cent of the sugar (Lehman, *c.f.* Barnes, 1918). The extraction efficiency of a 3-

roller Nahan mill at Pratabgarh (now Pratapgarh) farm was 59.7, 62.5 and 63.1 per cent in 1908-9, 1090-10 and 1910-11, crop seasons, respectively (Clarke *et al.*, 1912).

Table 14.2: Juice extraction by Old Wooden Mill and 3-Roller Nahan Mill

A. Juice extraction by wooden mill —the *belna*

Location	Per cent juice extracted
Karnal	42
Sialkot	50
Gurdaspur	50
Hoshiarpur	50
Jullundur	46 - 58
	67 - 69@

@ Due to thick canes—*paunda* type.

B. Juice extraction by 3-roller Nahan Mill at Gurdaspur (Punjab)

Variety	Fibre per cent cane	Extraction per cent	Per cent sucrose extracted
Tereru	17.73	60.5	76.8
Dhaulu	20.60	56.6	81.2
Katha	23.57	56.7	79.8
Kansar	18.33	61.5	84.1
Teru	16.05	58.0	78.4
Kahu	13.54	65.4	80.5
Katha-kahu	19.38	65.2	80.1
Paunda	10.60	65.3	78.0

Source: Barnes, 1918.

However, further investigations led to development of three, four or five roller crushers which are comparatively more efficient and effective in juice extraction. Number of rollers are dependent on the availability of power. An average pair of animals, which can develop 1.25 kW is used for pulling only a three roller crusher. However, mechanical power operated crushers are available to match the prime mover with the rated output of 3.7, 5.5, 7.5, 11, 15 and 20 kW. The source of power are not limited to animals and engines only. Some crushers have been found deriving power from water falls wherein water is being channelised to drive a turbine which finally drives the crusher at a rated speed. One such crusher was found installed on foot hills of Sri Lanka (Srivastava, 1990).

Manually operated sugarcane crushers are used for crushing small quantities of cane. They have two horizontal rollers. Their capacity is hardly about 10-15 kg/h and juice extraction is low (about 40%), mainly due to the limitation of force which can be exerted by hands. Because of the low extraction, the cane is passed 2-3 times to get 50-55 per cent extraction. The rollers are usually made of cast iron.

Animal operated sugarcane crushers are widely used all over the country where sugarcane is grown. They have usually 3 vertical rollers connected with each other with spur gears. The king roller which drives the other two rollers, derives power from a pair of animals through a 4-5 m long wooden beam. At a normal speed of animals between 2-3 kmph, the rollers rotate at a speed of 1.5 to 1.9 rpm. The juice extraction efficiency of such crushers is between 55-60 per cent and the capacity is about 125 kg of cane per hour. The rollers are also made of cast iron.

An improved bullock drawn crusher (Fig. 14.2) has been developed by M/s Kishan Krishi Yantra Udyog, Kanpur (Tromp, 1983). The project was financed by the Khadi and Village

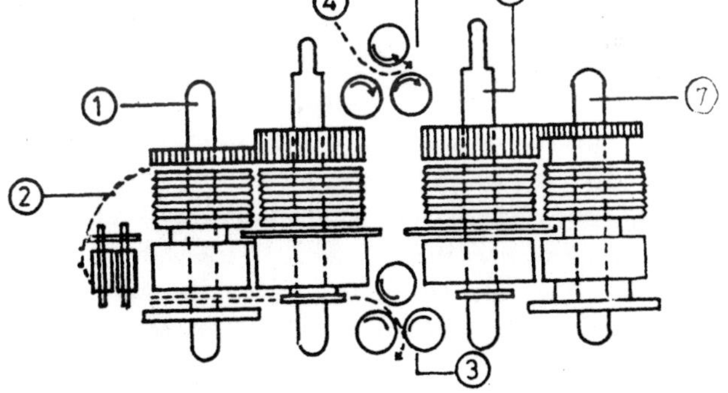

Fig. 14.2: Circumferential Grooves
1. Crushing Roller. 2. Cane Trough. 3. Final Bagasse. 4. Cane. 5. Prepared Cane. 6. King Roller. 7. Extraction Roller.

Industries Commission, India. It is claimed that the modified crusher yields 4 per cent higher extraction over the existing ones. In this design, the rollers are greater in length and are divided into two parts. While the upper 20 cm long upper part contains horizontal grooving, the 15 cm lower part is provided with vertical grooves. Thus, the upper portion of the rollers helps in cane preparation, *i.e.,* partial shredding of the cane and in the lower portion extraction of juice takes place.

During operation cane is fed in the upper portion. It gets shredded and at a discharge end it is turned and guided back towards the feeding end of the lower portion of the rollers where it is finally crushed and the juice is extracted. The roller shaft is provided with taper roller bearings which help in reducing the force requirement. At an output of 150 kg/h, the juice extraction has been claimed to be 68 per cent.

Power operated crushers are also very popular in the country. In recent years, the demand of power operated crushers has been comparatively more than the animal drawn crushers. They are run either by an electric motor of 3.75-7.5 kW range or even more by stationary oil engines. They have either vertical or horizontal rollers. Horizontal roller type power crushers are relatively more common. Their extraction efficiency is about 60-65 per cent (at 12.5% fibre content) and crushing capacity about 350-700 kg/h, depending upon the size.

Roller Grooving

Roller grooving is provided for many reasons *e.g.,* for better grip of cane, for help in breaking the cane into pieces and to guide extraction juice to trickle down. In case of bullock drawn crushers with vertical rollers, grooves are normally provided along the length of the rollers at intervals of about 8-10 mm. These grooves primarily help in increasing friction. In a few designs, grooves both along and across the axis of rollers are also provided. On horizontal roller type power crushers, grooves across the axis and both across and along the axis are more common. In this case, the grooves besides helping in better grip also serve, to some extent, as juice grooves.

In sugar mills where horizontal rollers are used, the following types of roller groovings are employed.

Circumferential Grooves

It is formed by grooving the roller with notches in complete circle (Hugot, 1979). The grooves are made perpendicular to the axis of the roller. The grooves are in the shape of an isoceles triangle with an apical angle of about 45-60° (Fig. 14.3. and Fig. 14.4). It is described in terms of h and p. Grooving of 10 × 12 means 10 mm deep and 12 mm pitch. The groovings of 3 roller mills are so arranged that the teeth of top roller meshes with the grooves of the two bottom rollers. This type of grooving is more popular.

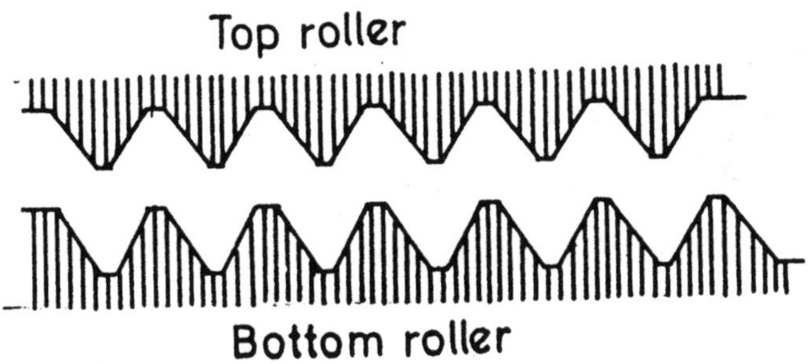

Fig. 14.3: Details of Circumferential Grooves

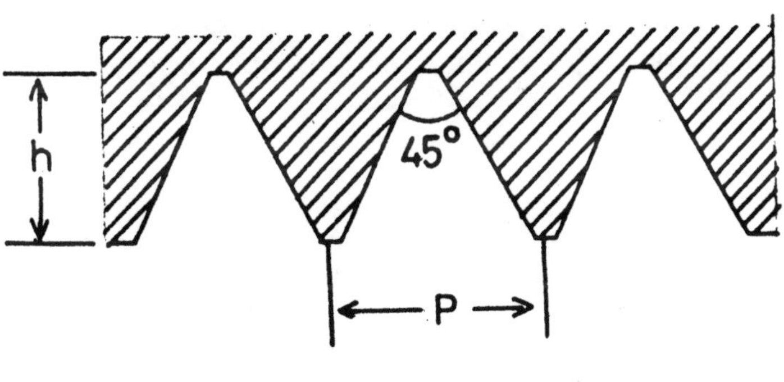

Fig. 14.4: Messachaert Grooves on the Bottom Roller

Messachaert Grooves

This is also called as juice groove. It was invented and applied in Hawaii with the objective of improving extraction. After every third or fourth tooth, deep grooves are provided on roller surface by either eliminating a tooth and locating Messachaert in the position of the tooth eliminated or by leaving all the teeth and cutting the Messachaert between two teeth as shown in Fig. 14.5. They have now become very popular.

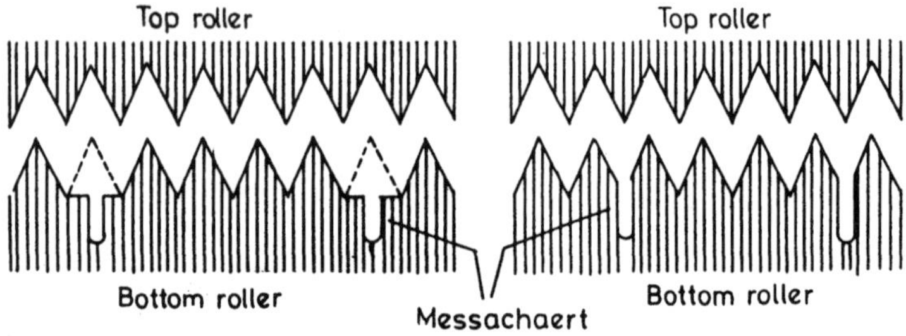

Fig. 14.5: Action of Messachaert Grooves in the AB Zone of the Crusher

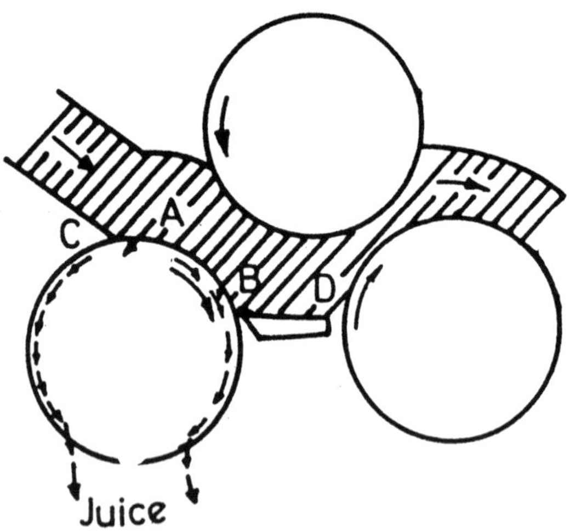

Fig. 14.6: Chevron Grooves on the Roller

As shown in Fig. 14.6 in the zone AB, the bagasse undergoes maximum pressure and the juice is extracted. This juice has only two ways to escape either forward with the roller, or backward in the direction of point C. In either direction it encounters a thick mat of bagasse. Due to provision of Messachaert groove juice can find an immediate way out in the circumferential grooving and can drain without hinderance from any part of the roller.

Chevrons Grooves

Chevrons, unlike Messachaerts, are provided with the sole objective of improving the feeding of the bagasse. In this case notches are cut in the teeth, in the form of a helix from the centre to one end of the roller (Fig. 14.7). The angle of the Chevron with a longitudinal line on the roller varies from 10-20° and generally 18° on big rollers, the pitch is normally 20 mm.

Kay Grooves

This type of grooving (Fig. 14.7) is also provided in the top roller for better feeding of the bagasse without any detrimental effect on the continuity of the pressure. The system consists of cutting helicoidal grooves of rectangular section of a depth greater than that of the circumferential grooving.

In operation, these grooves get filled with bagasse and the pads of bagasse so formed facilitate the feeding of the entering bagasse.

Effect of Diameter and Peripheral Speed on Juice Extraction

Roller Diameter

Roller diameter and peripheral speed of rollers are important from the point of view of juice extraction and capacity of the crusher. In small crushers, the variation in size of crushers in different designs is not much, but in factory crushers large diameter (75-100 cm) crushers are commonly used.

Some recommended diameters and face lengths of different

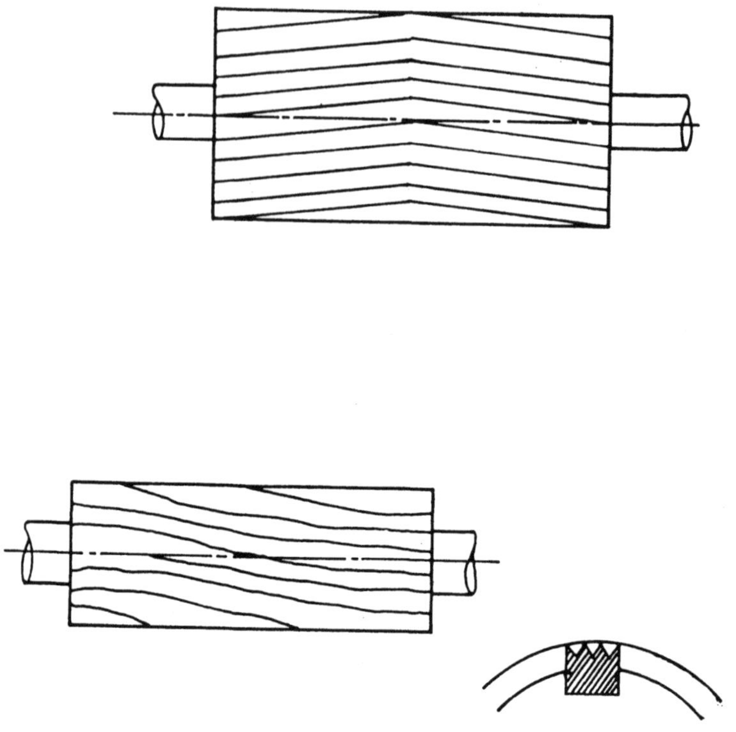

Fig. 14.7: Kay Grooving on the Roller

rollers used in animal and small power operated crushers (Tromp, 1973a; 1973b) are given below:

(a) *King Roller*

(i) diameter -212, 224, 236, 250, 265, 280, 300 and 315 mm
(ii) face length—150, 165, 212, 224, 250, 280, 315 and 355 mm

(b) *Crushing Roller*

(i) diameter—125, 150, 165, 212, 224, 236, 250 and 265 mm
(ii) face length—150, 165, 190, 212, 224, 250, 315 and 355 mm

(c) *Extraction Roller*

(i) diameter—212, 224, 236, 250, 265, 300 and 315 mm
(ii) face length—150, 165, 212, 224, 250, 280, 315 and 355 mm

The recommended diameters of steel bars used for the axle of these rollers are 50, 63, 65, 71, 75, 90 and 110 mm.

The most common sizes of rollers used in animal drawn crushers are: king 224 mm, crushing 165 mm and extraction 224 mm diameters. In an experiment with three sizes of king and extraction roller diameters (225, 175 and 125 mm), it was found that the effect of king and extraction roller diameters, gap between the rollers and speeds are highly significant with regards to juice extraction, crushing capacity and power requirement (Srivastava, 1986). The bigger diameter of king and extraction rollers yield significantly higher extraction and crushing capacity over the ones with smaller diameter. However, there is not much effect of diameter of the crushing roller on juice extraction, crushing capacity and the power requirement.

Studies conducted on factory crushers (Hugot, 1979; Meade and Chen, 1977; Murry, 1960; Murry and Holt, 1962; Shaw, 1971) have shown that the extraction is not dependent on diameter at the same surface speed in the range of 6 m/min to 18 m/min and same roller load. The torque is almost independent of the speed. This indicates that the mill performance should be compared at the same surface speed. On this basis, it appears that the reabsorption factor is independent of roller diameter. It has also been observed that the energy absorbed per tonne of cane crushed is likely to be inversely proportional to the roller diameter and that large rollers are more efficient.

Peripheral Speed

Peripheral speed of the crushers is important, both from the point of view of extraction and crushing efficiency. At lower speed, the cane is compressed for a larger time as a result juice gets relatively more time to trickle down and yet, at slower speeds, the crushing capacity is low. Chances of reabsorption of the juice, specially in case of crusher with vertical rollers is also more. In a study wherein animal drawn crusher was operated by varying speed electric motor, it was observed that at a peripheral speed of 2 m/min (2.6 rpm) highest extraction percentage was observed at all the gap settings (ranging from 0.5-2.0 mm) (Pandya, 1962). Srivastava and Agrawal (1985a, 1985b) and Srivastava (1986) observed that in the speed range of 1.5 to 14 rpm juice extraction decreased with the increase in speed of rollers. This was mainly due to two reasons. Firstly, at higher speeds, the cane was pressed for a relatively shorter time resulting in more juice being left in the bagasse. Secondly, as the rate of crushing increased at higher speeds, more juice per unit time was extracted from the canes being fed. However, this juice did not get enough time to trickle down and a substantial amount of juice, which remained adhering around the roller surface, got partly reabsorbed by the bagasse. The peripheral speed in the range of 1.06 to 2.1 m/min (1.5-3 rpm) appears to be desirable both from extraction and capacity point of view.

In factory milling, where there is large variation in roller diameter from one to another, expression of speed in terms of peripheral speed is rather more convenient (De Boer, 1973; Haines and Huges, 1962; Hugot, 1979; Shaw, 1971; Tromp, 1957) and useful than the rotational speed. It has been observed that higher the peripheral speed, the more difficult will be for the juice to escape against the movement of the bagasse and greater will be the amount of juice entrapped in the compressed bagasse where its permeability drops to practically zero. There is also difficulty in extraction due to the film of juice adhering to the roller and partly thrown towards the bagasse.

Several formulae have been evolved to correlate speed with diameter of the roller (Hugot, 1979; Tromp, 1846) and the one which is being used commonly in general practice by the factories is:

$$VE = 7.5 + 7.5\ D$$

and $$ne = 2.4 + 2.4/D$$

where,

VE = economic peripheral speed m/min
ne = economical rotational speed in rpm
D = diameter of the roller in metre

The rotational and peripheral speeds of factory crushers are generally around 5 rpm or 15 m/min.

Effect of Number of Rollers on Juice Extraction

Juice extraction increases with the increase in number of crushings as also with the number of rollers. Devdas (1986) found that when the number of rollers increased from 3 to 4, the juice extraction percentage increased by about 10 per cent but the capacity is reduced by about 14 per cent. As the number of rollers increased the power requirement also increased.

Srivastava (1986) and Srivastava and Agrawal (1988) found that the number of rollers had significant effect on extraction, crushing capacity and power requirement. A three roller crusher (having king, extraction and crushing roller) gives 2.7 per cent higher extraction, 30 per cent higher crushing capacity and requires about 5 per cent more power as compared to a two roller crusher (having only king and extraction rollers). A prediction equation has also been developed which shows that as the number of rollers (extraction) increases the juice extraction and power requirement also increases. However, when the number of extraction rollers is increased, the capacity is reduced.

Hugot (1979) has also mentioned that the extraction increases with the increase in number of rollers, as shown in Table 14.3.

Table 14.3: Extraction with Different Types of Crushers

Crusher	Extraction % absolute juice
Fulton (2 rollers)	35 - 55
Double crusher (4 rollers)	60 - 70
Fulton crusher followed by mill (5 rollers)	70 - 83

Effect of Gap Between the Rollers

On small crushers the gap setting between king and crushing roller is kept 3-5 mm and between king and extraction roller 0.25 to 1.75 mm. The most common setting used on animal drawn crushers is that of 3.125 and 0.75 mm, respectively.

Gap between the rollers (king and extraction) is important from the point of view of juice extraction and power requirement. Lower the gap higher is the juice extraction and also the power requirement. However, the gap between king and crushing rollers does not have any significant effect on extraction, crushing capacity and the power requirement (Singh, 1963; Singh and Arya, 1972; Srivastava and Agrawal, 1985a, 1985b; Srivastava, 1986).

For adjusting the gap between the rollers two screws are normally provided on animal drawn crushers, one at the top and another at the bottom. By loosening or tightening these screws, the position of bush blocks of extraction and crushing rollers are moved in and out. Although the arrangement looks crude, but in actual position when the cane is fed and the rollers are under pressure, the maintenance of gap is quite satisfactory.

Effect of Orientation of Rollers on Juice Extraction

Studies have shown that a horizontal roller crusher yields more juice extraction than the one with vertical rollers. Devdas (1986) found two per cent and Srivastava (1986) 3.25 per cent increase

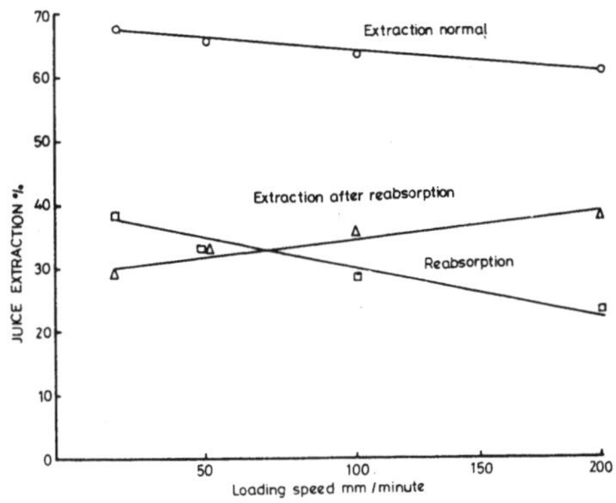

Fig. 14.8: Effect of Loading in Speed on Extraction of Juice

in juice extraction with horizontal rollers in comparison to vertical rollers.

In a vertical roller crusher, more than one cane is fed at a time in vertical plane. As the crushing commences, the juice extracted from the top cane falls on the bagasse of the second cane which is moving and expanding as it passes beyond the work opening. In the process, a part of the juice is reabsorbed and the extraction percentage is reduced (Srivastava and Agrawal, 1985a; 1985b) and Srivastava (1986).

Using Instron Universal Testing Machine it is observed that about 23 to 38 per cent of the extracted juice is reabsorbed (Fig. 14.8) by the bagasse if the juice is not removed before the relaxation process takes place. The reabsorption is comparatively slower when the load was released at higher speeds as the bagasse got less time for relaxation (Srivastava, 1986).

Maintenance of Rollers

When the rollers are in constant use, there is more wear and tear of the surface in the middle portion of the roller. The gap adjustment, therefore, becomes difficult. This calls for turning the roller to restore its shape to proper cylindrical form.

If the roller is badly worn out it can be reshelled by breaking off the outer cast iron shell and recasting a new shell.

Power Requirement

Extraction of juice from cane involves three steps:

(a) breaking up of the cane stalk to separate the cane cells from one another,
(b) expressing juice from the cells, and
(c) forcing the expressed juice out through the compressed mass.

In order to achieve the above objectives in small crushers, power is required/spent for the following:

(i) in the compression of cane or bagasse,
(ii) in friction between shaft and bearings of rollers,
(iii) absorbed in the gearings, and
(iv) consumed in the belt drive system, in case of power operated crushers and in driving the intermediate carriers in case of factory milling.

Besides, in case of sugar factories, additional power is consumed in the following:

(v) friction between bagasse and trash plate,
(vi) friction of scrappers and toe of the trash plate against rollers.

Studies have shown that the power requirement increases with the increase of speed (Bullock, 1956; 1957; Bullock and Murry, 1959; Crawford, 1959; Deerr, 1946; Holt, 1963; Hugot, 1979; Jenkins, 1950; 1957; Lekhy, 1937; Mury, 1959a, 1959b; Srivastava, 1986; Srivastava and Agrawal, 1985a; 1985b; 1988; Tromp, 1973a). It is also inversely proportional to the gap size. Power is directly proportional to the compression ratio (Singh, 1963; Singh and Arya, 1972; Srivastava, 1986; Srivastava and Agrawal, 1985a; 1985b; 1988), rind hardness and fibre content of the cane.

On animal drawn crushers, mostly wooden bearings are used for roller shafts, presumably due to very slow speed and ease of replacement in the villages. However, some of the power crushers are provided with gun metal bushings. Srivastava (1986) observed that in animal drawn crushers different types of bearings, namely, wooden bearings, gun metal bushings, nylon bushing and ball bearing do not have any significant difference in power requirement at slower speeds of operation (1.5 to 3 rpm). However, at higher speeds of 5 rpm and above, there is some difference in power requirement. At higher speeds, the power requirement is lowest in case of ball bearing followed by gun metal bearings. Juice extraction is not affected by the type of bearing.

Preparation of Cane Material

In case of animal drawn and small power crushers no preparation of cane is done before crushing. Cane is fed in as it is. But, in factory milling, preparation of cane material is done before crushing with a view to get uniform and homogeneous mat of bagasse for feeding into the mills with less voids. This is accomplished in following ways:

(i) by revolving cane knives that cut cane into chips but extract no juice,
(ii) by shredders that tear the cane into shreds but extract no juice,
(iii) by crushers that break and crush the structure of the cane, extracting large proportions of the juice,
(iv) by combination of any two or three methods mentioned above.

Bullock (1956) and Murry and Shann (1959) reported that extraction is a function of preparation and that a definite relationship exists between preparation, compression and the extraction. This shows that the response of extraction to the changes in fineness of preparation differs at different compression ratio. Their study also indicates that within the practical compression ratio range, increasing fineness of preparation shows no drop in extraction or in ultimate rate, i.e., no 'optimum' preparation could be identified. However, it is felt that very fine preparation like saw dust may lead to a drop in extraction.

Srivastava (1986) studied the effect of cane preparation on juice extraction capacity and power requirement in animal operated crusher using five types of materials, namely, whole canes, whole canes with nodes smashed, cane split into two halves, cane split into four pieces and rind removed. It is observed that scraping of rind and splitting of cane reduced the draft (power) requirement, as well as the juice extraction. Only smashing of nodes and feeding the whole cane gives juice extraction *at par* with whole cane and requires less draft. In case of samples with scraped rind, the draft requirement is minimum but the juice extraction is quite low as compression ratio is reduced due to reduction in diameter of the samples.

Co-efficient of Friction

Bullock and Murry (1958) investigated the co-efficient of friction of bagasse on iron surface. It varies widely under different conditions. These findings indicate:

(a) The co-efficient of friction decreases with increasing rubbing speeds but the increase is less rapid with dry fibre than with fresh cane.

(b) Fibrated preparation gives higher co-efficient of friction than coarse preparation but at higher pressures and in fresh canes the difference is less marked.

(c) The groove surface gives higher apparent co-efficient of friction than the smooth surface, the difference is more pronounced at higher pressures and also in case of dry fibre.

(d) Dry fibre gives higher co-efficient of friction than fresh cane but is less marked at low speed, coarse preparation, and smooth surface. It is more pronounced at higher pressures.

(e) The real co-efficient of friction decreases with increasing pressure but due to packing against the surface, the apparent co-efficient of friction is relatively higher at higher pressure than at lower pressures.

The value of co-efficient of friction of fresh canes (coarse) ranged from 0.13 to 0.34 at pressure on the rollers ranging from 1.4 to 6.0 kg/cm^2. Higher the co-efficient of friction more will be the pressure exerted on the cane resulting into higher extraction.

Capacity of the Crusher

The capacity of the sugarcane crusher is dependent on the following:

(i) Diameter of the roller
(ii) Peripheral speed of the roller
(iii) Gap between the rollers or work opening
(iv) Co-efficient of friction between the cane and the roller surface
(v) Thickness of cane-mat entering the roller
(vi) Number of cane/width of mat being fed at a time.

In cane mills, the tensile strength of the feed blanket, the ratio between the transverse and longitudinal pressures and pressure volume relationship, which is chiefly the function of cane preparation, also contribute towards the capacity of a sugarcane mill. It is observed that there is a significant change in co-efficient of friction in cane preparations. The tensile strength of coarse preparation is less which results in the reduction of co-efficient of friction. The dynamic longitudinal pressure becomes significantly higher because of the lower permeability of the coarse preparation (Bullock, 1957; 1958a; 1958b; 1958 c, Bullock and Murry, 1958).

Studies conducted later show that the capacity of the crusher is directly proportional to the gap between the rollers, their speed and the feeding rate. Higher the gap, speed and feeding rate, higher will be the capacity of the crusher (Srivastava, 1986; Srivastava and Agrawal, 1985a; 1958b).

Safety Devices on Sugarcane Crushers

Some accidents have been reported while operating sugarcane crushers. They are mainly due to entanglement of hand in the

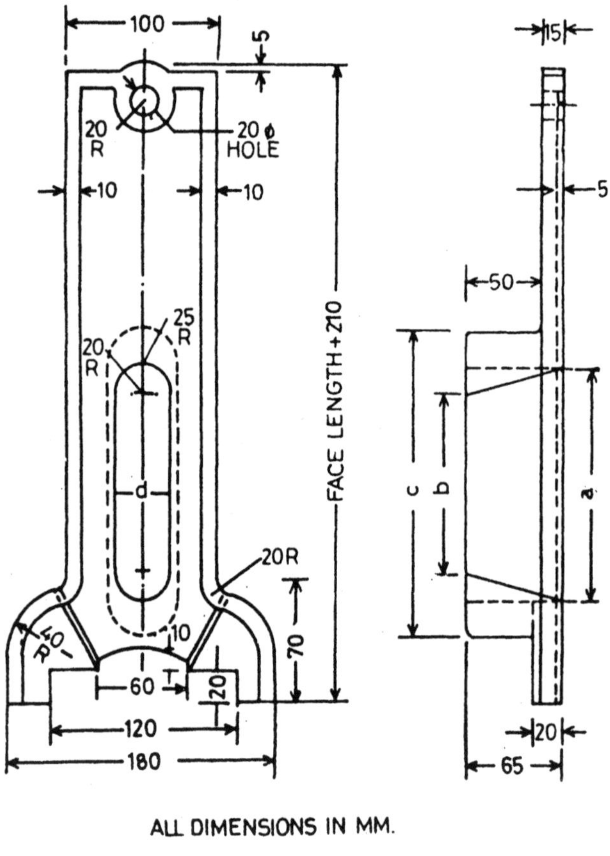

Fig. 14.9: Details of Feed Plate

rollers while feeding canes. Provision of a feed plate over the rollers avoids this type of accidents (Tromp, 1980). The dimensions of the feed plate (Fig. 14.9) have been specified according to the roller size (Table 14.4). These dimensions have been standardised by the Indian Standard Institution, New Delhi.

Table 14.4: Critical Dimensions of Feed Plate for Different Face Length of Rollers of Sugarcane Crushers

Face length of rollers (mm)	a	b		d	Total length of feed plate (mm)
150	140	110	150	40	360
165	140	110	150	40	375
190	140	110	150	40	400
212	180	150	200	40	422
224	180	150	200	40	435
250	210	180	240	40	460
280	210	180	240	40	490
315	260	230	290	40	525
355	260	230	290	40	565

Dimensions other than a, b, c and d in Fig. 14.9 are recommedatory.

Prediction Equations for Calculating Juice Extraction, Capacity and Power Requirement

Prediction equations have been developed to estimate the values of juice extraction, crushing capacity and power requirement (Srivastava, 1986). Estimated values are very close to the observed values and the variations, in most of the cases, are less than 1.5 per cent. These prediction equations give reliable estimates within the limits that the values of independent variables are rational and are neither zero nor arbitrary. These equations are given below:

For animal drawn sugarcane crusher

$$Je = 90.75 - 3.128\ x_1 + 0.368\ x_2 - 3.05\ x_3 - 0.383\ x_4 + 2.579\ x_5$$

$$C = -137.415 + 2.77\, x_1 + 2.999\, x_2 - 2.856\, x_3 + 42.211\, x_4 + 32.149\, x_5$$

$$P = 2.44 - 0.218\, x_1 + 0.02\, x_2 + 0.137\, x_3 + 0.227\, x_4 + 0.047\, x_5$$

For power operated sugarcane crusher

$$Je = 74.32 - 3.2\, x_1 + 0.375\, x_2 - 1.35\, x_3 - 0.279\, x_4 - 0.593\, x_5$$

$$C = 3.125 + 12.389\, x_1 + 3.3\, x_2 - 11.797\, x_3 + 43.59\, x_4 + 44.389\, x_5$$

$$P = 0.101 = 0.15\, x_1 - 0.01\, x_2 - 0.0077\, x_3 - 0.274\, x_4 + 0.275\, x_5$$

For experimental sugarcane crusher

$$Je = 58.5 - 3.61\, x_1 + 0.38\, x_2 - 1.53\, x_3 - 0.66\, x_4 - 0.69\, x_5 + 0.0852\, x_6$$

$$C = -327.423 + 7.573\, x_1 + 1.0\, x_2 + 14.817\, x_3 + 36.786\, x_4 + 15.076\, x_5 + 0.322\, x_6$$

$$P = 4.847 - 0.159\, x_1 - 0.009\, x_2 - 0.381\, x_3 + 0.296\, x_4 + 0.359\, x_5 + 0.005\, x_6$$

where,

Je : Juice extraction (%)
C : Capacity (kg/ha)
P : Power (kW)
x_1 : Gap (king and extraction roller, mm)
x_2 : Diameter of cane (mm)
x_3 : Fibre content of cane (%)
x_4 : Feeding rate (number of canes)
x_5 : Number of rollers
x_6 : Diameter of rollers (mm)

Reduction in Juice Extraction

Since the juice extraction is very much dependent on the fibre content of the cane, any particular crusher at a particular setting will give different juice extraction depending upon the fibre content of the cane. In order to have uniformity, in comparing results, the juice extraction percentage is expressed at 12.5 per cent fibre. For calculating the reduced juice extraction at 12.5 per cent fibre, the following formula called the "Reduced Juice Extraction Mittal" is used.

$$E_{12.5} = 1 - \frac{0.125\ (1 - e)}{f}$$

where

E = Reduced juice extraction at 12.5 per cent fibre
e = Juice extraction obtained at fibre f
f = Fibre content of cane.

Crushers in Open Pan Sulphitation (OPS) Technology of *Khandsari* Making

The OPS technology employed is a mixture of traditional and modern techniques. For crushing, a new 4-6 roller hydraulic crusher with two sets of knives and capacities of 4 tonnes/h has been introduced. Milling efficiencies of 75 per cent is achieved without inhibition. Small crushers had previously not used knife preparation although spring loading had been in use (McChesney, 1989). In the improved OPS process adopted by the Planning and Research and Action Institute (PARI), Lucknow and Applied Technology Development Association (ATDA), for crushing, a single stage screw extrusion device called *cane expeller,* has been developed so as to attain a milling efficiency of 85 per cent at capacities of 2-2.5 tonnes/h (Garg, 1979; Varma and Garg, 1987). As normally practised in sugar beet, a small scale diffuser technology has also been developed as a new cane extraction technology for small scale sugar processes. The first commercial diffuser installation was commissioned in 1962 in Tanzania (James, 1989).

REFERENCES

Anonymous (1982). *Indian Standard Specification for Sugarcane Crushers* (Second Revision) UDC 664.1.034.2: 621.927, IS: 1973-1981, Indian Standards Institution, New Delhi, p. 14.

Anonymous, (Anakapalle, India) (1989). *75 Years of Agricultural Research at Anakapalle ,1913-1988*. RARS, Anakapalle, A.P.A.U., Anakapalle, A.P., p. 400.

Anonymous, (1990). *Glimpses of Cane, Gur and Khandsari Industries*, Under KVIC, Directorate of *Gur* and Khandsari Industry, K.V.I.C., Gramodaya, Bombay, p. 46.

Barnes, A.C., (1974). *The Sugarcane*, Leonard Hill Books, Aylesbury, Bucks.

Barnes, J.H. (1918). Sugar and the sugarcane in the Gurdaspur district. *Bulletin No. 69* Agricultural Research Institute, Superintendent Govt. Printing, India, Calcutta, p. 100.

Bullock, K.S. (1956). The effect of speed, preparation etc. *Proc. Q S S C T Conference*, 22: 301.

Bullock, K.G. (1957a). An Investigation into the Crushing and Physical Properties of Sugarcane and Bagasse, Ph.D. Thesis, University of Queensland, Australia.

Bullock, K.G. (1957b). Self feeding characteristics of an experiment sugar mill. *Intern. Sugar J.*, 59: 48.

Bullock, K.G. (1958a). The capacity of sugarcane mills, Pt. I. *Sugar J.*, 22 (5): 11-17.

Bullock, K.G. (1958b). The capacity of sugarcane mills, Pt. II. Experiments on feeding a two roll mill. *Sugar J.*, 22(6): 26-29.

Bullock, K.G. (1958c). The capacity of sugarcane mills, Pt. III. Theoretical analysis of mill feeding. *Sugar J.*, 22(7): 13-20.

Bullock, K.G. and C.R. Murry (1958). Co-efficient of friction of bagasse on iron surface. *Intern Sugar J.*, 60(714): 162-154.

Bullock, K.G. and C.R. Murry (1956): Extraction as a function of roll load in the two roll experimental mill. *Proc. I.S.S.C.T.*, 10: 111-122.

Chang, Lie-Tien (1992). Cane sugar manufacture in ancient China. *International Sugar J.*, 94 (1123): 155-156.

Clarke, G., H.E. Annett, S.Z. Hussain, S.C. Banerji and N. Hussain (1912). Experiments on the cultivation of sugarcane at the Pratabgarh Experimental Station 1909-1911, Agricultural Research Institute Publication, Superintend, Govt. Printing, India, Calcutta, p. 29.

Davidson, L.G. (1958). Correlation of rind hardness and fibre in sugarcane. *Sugar J.*, 31(5): 5-11.

De Boer, A.T. (1973). Cane mill extraction the relation between juice extraction, fibre index, top pressure and circumferential roller velocity. *Intern. Sugar J.*, 75(898): 306-308.

Deerr, N. (1946). Power required for milling cane. *Intern. Sugar J.*, 48: 234.

Devdas, C.T. (1986). Evaluation of animal and power operated sugarcane crushers. *Annl. Rept. (1985), PHTS Scheme*, TNAU, Coimbatore, Tamil Nadu, India.

Garg, M.K. (1979). *Project Rep. and Feasibility Study of Appropriate Technology on Mini Sugar* (OPS Khandsari), ATDA.

Gazeeter of Oudh (1977-78). Vol. I, Reprinted in 1985, B.R. Publication Corporation, Delhi, p. 609.

Ghosh, M.K., S.V. Dixit and H.C. Goyal (1956). Some Tests on the Performance of Bullock Driven Crusher, B. Tech. thesis, Deptt. Agril Engg., I.I.T., Kharagpur, India.

Gupta, R.K. and U.C. Sood (1957). A Study of Performance of Bullock Drawn

Sugarcane Crusher, M. Tech. Thesis, Deptt. Agril. Engg., I.I.T., Kharagpur, India.

Haines, W.S. and R.H. Huges (1962). Investigations on extraction in commercial milling tasndem. *Proc. I.S.S.C.T.,* 11: 1013-1030.

Holt, J.E. (1963). The Prediction of Roll Loads in the Crushing of Prepared Sugarcane, M. Tech. Thesis, University of Queensland, Australia.

Hugot, E. (1979). *Hand Book of Cane Sugar Engineering,* Elsevier Sci. Publishing Co., Amsterdam, Oxford and New York.

Hulme, W. and R.P. Sanghi (1918). The improvement of indigenous methods of *gur* and sugar making in the United Provinces. *Bulletin No. 82,* Agricultural Research Institute, Superintendent, Govt. Printing, India, Calcutta, p. 22.

Humbert, R.P. (1968). *The Growing of Sugarcane,* Elsevier Sci. Pub. Co., Amsterdam, Oxford and New York.

James, A. (1989). New cane extraction technology for small scale factories. In: *Cane Sugar: The Small Scale Processing Option,* Proc. Jt. ITDG/IDS Conf., Sept. 10-11, 1987, Kapilnsky, R. (Ed), Intermediate Technology Publications, London, pp. 61-63.

Jenkins, G.H. (1950). A basis of assessing power requirements in cane milling. *Proc. I.S.S.C.T.* 9 (II): 154-155.

Jenkins, G.H. (1957). Pressure in the three roller mill. *Sugar J.,* 20 (2): 12.

Khanna, K.L., (1935). An improved instrument for testing rind hardness of sugarcane. *Agric. and Livestock in India,* 5 (II): 156-158.

Lekhy, R. (1937). Calculation of the driving power for cane crushing mills. *Intern. Sugar J.,* 39 (460): 137-141.

Martin, F.A. and B.T. Cocharan (1975). Sugarcane internode rind hardness. *Sugar Y Azucar* 70(2): 26-30.

McChesney, I. (1989). Improvements in open pan sulphitation technology. In: *Cane Sugar: The Small Scale Processing Option,* Proc. Jt. ITDG/IDS Conf., Sept., 10-11, 1987, Kapilnsky, R. (Ed), Intermediate Technology Publications, London, pp. 71-73.

Meade, G.P. and J.C.P. Chen (1977). *Cane Sugar Handbook,* John Wiley and Sons, New York.

Mungomery, R.W. (1936). The rind hardness machine and its use. *Cane Growers Bull.* 4: 51-53.

Murry, C.R. and D.S. Shann (1959). Effect of degree of preparation on the Queensland experimental mills. *Proc. I.S.S.C.T.,* 10: 122-131.

Murry, C.R. (1959 a). Mill torque as a function of roll loading. Pt. I. Preliminary considerations. *Intern. Sugar J.,* 61(722): 46-50.

Murry, G.R. (1959b). Mill torque as a function of roll loading. Pt. II. Development of dimensional analysis. *Intern. Sugar,* 61(723): 72-76.

Murry, C.R. (1960). The Mechanics of Crushing Prepared Sugarcane, Ph. D. thesis, University of Queensland, Australia.

Murry, C.R. and J.E. Holt (1962). The effect of roll diameter on the mechanics of crushing. *Proc. I.S.S.C.T.,* 11: 1074-1086.

Pandya, A.C. (1962). Performance of a bullock operated sugarcane crushers. *J. Agric. Engg. Res.,* 7(1): 3-7.

Puri, A.N. and T.S. Venkatraman (1929). An apparatus for testing rind hardness in sugarcane. *Proc. S.T.A.I.,* 1929: 1-17.

Ramana Rao, T.C., K.V. Bhagyalashmi and J.T. Rao (1979). *Indian Atlas of Sugarcane,* Sugarcane Breeding Institute, Coimbatore, India.

Shaw, M. (1971). Milling research in the University of Queensland. *Proc. I.S.S.C.T.*, 14: 1342-1359.

Singh, Y. (1963). Studies on Sugarcane Juice Extraction, M. Tech. thesis, Deptt. Agril Engg., I.I.T. Kharagpur, West Bengal, India.

Singh, Y. and S.V. Arya, (1972). Performance of sugarcane crusher. *J. Agric. Engg.*, 9(1): 1-10.

Skinner, J.C. (1974). Rind hardness and fibre content. *Proc. I.S.S.C.T.*, 15: 153-167.

Srivastava, A.C. (1990). *Report on Sugarcane Farm Machinery Research of the Sugarcane Research Institute, Sri Lanka: Its Current Status and Future Directions* (May 22- June 21, 1990). Project No. SRL/84/007, FAO of UN, Rome, Italy, pp. 54.

Srivastava, N.S.L. (1986). To Evolve the Optimum Parameters for the Design of Animal Drawn Sugarcane Crusher Under Dynamic Loading, Ph.D. thesis, Faculty of Engg., Bhopal University, Madhya Pradesh, India.

Srivastava, N.S.L. and D.D. Agrawal (1985a). Animal operated sugarcane crusher - Effect of different parameters on juice extraction. *Proc. I.S.A.E. SJC*, 3(111): 150-160.

Srivastava, N.S.L. and D.D. Agrawal (1985b). Research and development in sugarcane crushing for jaggery industry. *Proc. National Seminar cum Group Discussion on Jaggery Manufacture and Storage*, Indian Institute of Sugarcane Research, Lucknow, Dec. 18-19, pp. 72-79.

Srivastava, N.S.L. and D.D. Agrawal (1988). Prediction equations for juice extraction, capacity and power requirement of animal drawn sugarcane crusher under varying machine and crop conditions. *Indian J. Sugarcane Technol.*, 5(2): 112-117.

Tromp, L.A. (1946). *Machinery and Equipment for Cane Sugar Factory*, Norman Rodger, London.

Tromp, L.A. (1957). Milling studies 1. Influence of roller speed on No. I Mill performance. *Tech. Rept. No. 37*. Sugar Research Institute, Queensland, Australia (*Abst. Intern. Sugar J.*, 60: 170).

Tromp, L.A. (1968). Report of the 4th Technical Seminar on the Open Pan Sugar Manufacture, Sept., 24-26, PARI Publication No. 350. PARI, Lucknow, India.

Tromp, LA. (1973a). Specifications for Sugarcane Crusher, Indian Standard Institute, New Delhi.

Tromp, L.A. (1973b). IS 6983-1973. Specifications for Rollers and Axles for Sugarcane Crushers, Indian Standard Institute, New Delhi.

Tromp, L.A. (1980). Safety Device for Cane Crushers and Chaff Cutters. *Rept. of the Expert Committee Meeting* held at CIAE, Bhopal, Madhya Pradesh, India.

Tromp, L.A. (1983). *Ganna Kolhu me Sudhar, Gram Shilpa* (July-Sept.): 81-84. (in Hindi).

van Dillewijn, C. (1952) *The Botany of Sugarcane*. The Chronica Botanica Book Co., Waltham, Mass., USA, pp. 324.

Verma, M.C. and M.G. Garg (1987). Recent advances in the open pan sulphitation process of cane sugar manufacture. *Bharatiya Sugar Directory*: 258-268.

15

FURNACES

In *gur* manufacture, the juice boiling furnaces occupy a prominent place since clarification of the raw juice and its boiling so as to concentrate it to the required concentration is achieved in the furnace.

A simple juice boiling furnace consists of a burning chamber, with or without grate, and a flue channel at the end of which a chimney is provided for taking out flue gases. In single pan furnace the juice boiling pan is placed on the burning chamber, while in multiple furnaces pans are also placed on the flue channel as well to utilise the heat from the flue gases so as to effect fuel economy. The big multiple pan furnaces are generally more efficient than the single pan furnaces and have rather a lower fuel consumption per unit of the juice processed. But being more expensive, small and marginal farmers have necessarily to use the small furnaces in spite of their higher fuel consumption and relatively lower efficiency. The major drawback with most of the common furnaces is their excessive fuel consumption and their low index of heat utilisation. The necessity of extra fuel is especially felt when the canes of low fibre content are crushed.

FURNACES USED IN VARIOUS PARTS OF INDIA

Some of important furnaces used in various parts of our country are mentioned below:

Uttar Pradesh

A large number of different types of *gur* making furnaces are found in Uttar Pradesh. Each type of furnace is popular in a particular region of the State and accordingly derives its name. The following furnaces have been found to be most suitable for the manufacture of *gur* in the different regions:

Western Uttar Pradesh: Improved Meerut *Bel*
Central Uttar Pradesh: *Bilari Bel* No. 1 and 2
Eastern Uttar Pradesh: Improved South Bihar Furnace

The description of these is given below:

Desi Meerut Bel

This multiple furnace is used for manufacturing either *gur* or *rab* mostly in Meerut and adjoining districts. It consists of two round bottomed pans, feed and air holes and a chimney. Sometimes a third rectangular pan 2.67 m long, 53.34 cm wide and 30.34 cm deep is also placed over the flue passage between the first pan and the chimney so as to increase the capacity and to economise the fuel consumption. This furnace can concentrate about 149.2 kg of cold juice per hour, to *gur*.

Improved Meerut Bel

To improve the working of the *Desi Meerut bel*, the following modifications were made in the furnace, flue and chimney of the *bel*:

A. *To reduce fuel consumption and to increase juice boiling capacity:*

(a) A grate was introduced in the furnace for proper burning of bagasse.
(b) Arrangement of preheating air was made for raising the temperature of the furnace.
(c) The height of the chimney was increased from 3.05 m to 4.27 m for improving the draught.
(d) A rectangular gutter pan 2.45 m × 45.72 cm × 30.48 cm was added between the first pan and the chimney to utilise some of the heat escaping out with the flue gases.

B. *To Prevent Caramelisation*

(a) One removal *parcha* (striking pan) 76.2 cm dia × 30.5

cm deep was added between the two round pans of the *bel* to facilitate the striking of the finished product.

(b) The improved Meerut *bel* can process 559.5 kg of hot juice per hour i.e., 447.6 kg of cold juice per hour for *rab* (massecuite) or 373.0 kg cold juice per hour for *gur* manufacture. The furnace requires about 1268 kg of bagasse for boiling 3730 kg of juice.

Bilari Bel No. 1

This multiple furnace has been specially designed for small cane growers. The main features of this *bel* are a gutter pan, a provision for tilting the finished product in the pan, a grate for complete burning of bagasse and an air preheating device.

The *bel* consists of one rectangular gutter pan and two flat bottomed pans fixed on mud ovens over a slanting flue pipe. The gutter pan is 1.37 m long × 45.72 cm wide × 30.48 cm deep. The bottom of the pan is made of 0.3 cm mild steel sheet and its sides are of 0.15 cm thick sheet. The remaining two pans of the *bel* are of equal size and shape. They are flat bottomed, having a diameter of 114.3 cm at the top and 99.1 cm at the bottom and 30.48 cm in depth. The first round pan, next to the gutter pan, is called the *nikhar* pan and has an outlet pipe attached to its bottom for discharging clear juice into the second pan. The second pan called the finishing pan has a hinge on one side near the bottom and a ring is fixed on the rim opposite to the hinge to fasten the hook for tilting the pan, by means of a rope and pully at the time of making a 'strike'.

For utilising the heat of the flue gases a pre-heating passage 25.4 cm wide and about 2.4 cm deep has been made below the flue passage. The pre-heating air passage has two openings, one on each side of the gutter pan wall and the other at the bottom of the grate. The furnace has a rectangular grate 55.9 sq cm and in order that the flame passing the flue may strike the bottom of the pans, baffles (*thokar*) have been provided under all the three pans. A 4.27 m high chimney yield enough force to draw hot air easily into the furnace.

The boiling capacity of the Bilari *bel* No. 1 is 136.9 kg of juice per hour and its fuel consumption is 33.5 per cent on the quantity of juice boiled.

The capacity as well as the fuel consumption of a *bel* are greatly affected by the velocity and direction of the wind. But in this *bel,* the air for consumption does not come to the furnace directly from the atmosphere, but through the pre-heating channel below the flue. Since, there is no other entrance for the air, the functioning of the *bel* is independent of the wind direction and works at rather the same rate under all conditions. Also, the tilting arrangement of the striking pan ensures that there is no burning or caramelisation of *gur* at the time of striking. This increases the juice boiling efficiency of the *bel* and reduces the chances of spoiling the colour and taste of the *gur* during the critical stage of concentration.

Bilari Bel No. 2

This is the second *bel* of the series designed by the *gur* and Khandsari Research Scheme, at the Indian Institute of Sugar Technology, Kanpur (now the National Sugar Institute). In design this *bel* is similar to the Bilari *bel* No. 1 but relatively bigger in size. The rectangular gutter pan of the *bel* is 244 cm long, 60.96 cm wide and 30.48 cm deep. The dimensions at the *nikhar* pan have also been similarly increased. Its top and bottom diameters are kept 1.45 m × 1.29 m × 30.48 cm. It has an out let pipe attached to the bottom for discharging clear juice into the second (finishing) pan (the size being similar to that of the Bilari *bel* No. 1). It has a hinge and ring for tilting at the time of 'strike'. The other constructional details are similar to Bilari *bel* No. 1. The juice boiling capacity of this *bel* is 143.23 kg per hr and fuel consumption is about 34 per cent on juice.

Bihar

Improved South Bihar Furnace

This is an improved type of single pan furnace, designed by the Department of Agriculture, Bihar. The main hearth of the furnace is divided into two parts, viz., the burning chamber and the ash pit. The conical burning chamber has a diameter of 91.44 cm at the top and 1.37 m at the bottom with a depth of 76.20 cm from the ground level. The ash pit, 91.44 cm deep with bottom diameter of 91.44 cm, is situated just below the burning chamber. A grate made-up of a few iron rods is placed at the top of the ash pit having a diameter of 38.10 cm. The flue passage of the furnace is 2.44 m long and ends in a brick built chimney which is 3.35 m in height. The inner cross section of the chimney which is 25.4 cm × 25.4 cm. The flue from the furnace passes through the pipe lines laid in the flue passage. The flue lines are surrounded by a mud jacket, extending from the furnace to the chimney, for the passage of the cold air. Due to difference in pressure the cold air is sucked automatically into the burning chamber through a small aperture in the mud jacket near the base of the chimney. The air, while passing through the mud jacket, comes into contact with the flue pipe and picks up heat. The preheated air then passes down the ash pit and enters into the burning chamber through the grate.

The feed hole of the furnace is 20.32 cm × 20.32 cm in size and is flushed with the ground, a little away from the burning chamber, a few centimetre above the grate. The feed hole is fitted at the mouth with a vertically situated and automatically closing iron shutter. This arrangement facilitates the fuel feeding without any difficulty and at the same time prevents the entry of cold air into the burning chamber through the feed hole.

This type of furnace is self-sufficient in respect of fuel. It consumes about 35 metric tonnes of sun dried bagasse for every 100 tonnes of juice boiled. The rates of juice boiled and *gur* produced per hour in such furnaces are 44.76 kg and 7.5 kg, respectively.

In order to increase the efficiency of the furnace still further, the flue pipe mentioned above may be made of a few old kerosene oil tins (with the bottom removed), the tin being telescoped one into the other and the joints plastered with mud. The opening of the flue into the burning chamber is reduced by means of a few bricks and mud, to an area of 152.4 sq. cm so as to ensure a proper draught. The fuel consumption in such furnaces has thus come down to about 31 t for 100 t of juice boiled. The rate of boiling 46 kg of juice per hour to produce 9 kg of *gur*.

The use of triple metallic flue pipes instead of a single one showed a considerably higher efficiency on account of the greater heat transferred. The fuel consumption in such a furnace has come down to 29.5 t per 100 t juice boiled with an increase in boiling capacity to 69.0 kg of juice per hour producing 11.56 kg of *gur*.

The juice boiling pan (of mild steel 1.08 cm thick at the bottom and 0.32 cm at the sides) for the furnace should be flat bottomed instead of shallow concave bottomed—the one generally used by the cultivators. The top and bottom diameter of the pan are 132 cm and 101 cm with a depth of 17.78 cm. Such a pan can boil 111.9 kg of juice, but it will be desirable to boil 74.6 kg juice per charge for obtaining a quality product.

Lucknow Bel

Gur and *Khandsari* Research Scheme located at Lucknow, designed and developed a double pan furnace for small scale *gur* manufacturers (Fig. 15.1). The striking pan is circular having flat bottom with a sloping side. The diameter of the top and bottom of this pan are 145 cm and 135 cm, respectively with a depth of 25 cm. The other pan is rectangular in shape and is placed at a relatively higher level than the striking pan. It is usually heated by the hot flue gases. The size of the rectangular pan is 262 cm long and 80 cm wide and 30 cm deep, with a provision of outlet. Below the striking pan, in the front, is a 'step grate' of 23 cm wide and 45 cm in length. The air required for combustion of the bagasse is drawn through the hole of flue gases pipe of 10 cm diameter placed in the furnace. One end of this pipe opens in the ash pit and the other end opens in the sides of chimney 2.7 m in height. A door is provided below the grate for cleaning the ash pit at the desired intervals. It is normally opened once a day to take out the ash and closed with mud plaster thereafter. The cold air passing through flue pipe gets sufficiently heated and enables the burning of semi-dried bagasse. The second pan (or gutter pan) which is placed over the passage of flue gases is filled with cane juice for partial clarification to be used as next charge.

The distinctive feature of this furnace is the incorporation of a 'step grate' which helps in complete combustion of the bagasse. The arrangement for preheating of the air in closed

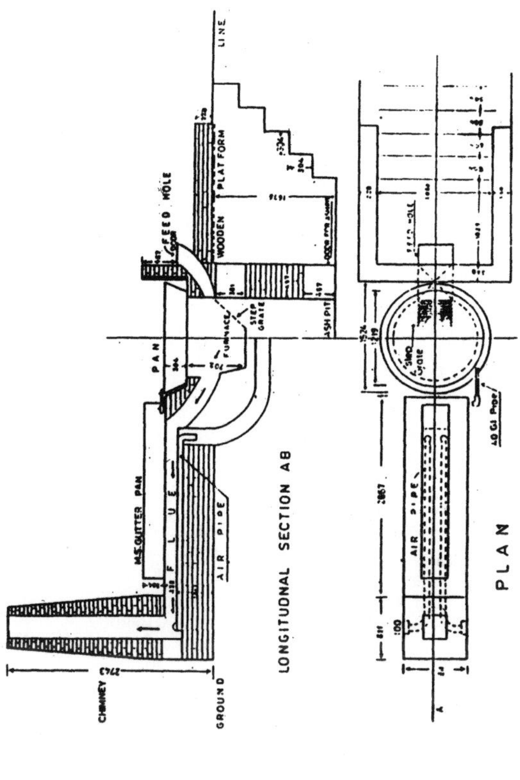

Fig. 15.1: Details of Lucknow Bel

pipe is very efficient because the air attends a high temperature almost equal to that of the flue gases. These improvements have resulted in low consumption of the fuel, high rate and increased efficiency of boiling with respect to the percentage of sugar recovered in *gur* from the total sugar present in the juice.

A comparative assessment of the Lucknow *bel* with *Bilari Bel* No. 1 is given in Table 15.1.

Table 15.1: Comparison of Lucknow Bel with Bilari Bel No. 1 for *Gur*-making

Particular	Lucknow *bel*	Bilari *bel* No. 1
Juice boiled (kg h^{-1})	102.2	127.2
Fuel efficiency (Bagasse consumed % juice)	40.0	33.5
Efficiency (pol in juice per cent pol in *gur*)	89.3	88.4

Source: Anonymous, 1955.

Though the Bilari *bel* is somewhat superior to the Lucknow *bel* with respect to juice boiled per unit time and fuel consumption, the efficiency of the latter is relatively more than the former. Furthermore, the Lucknow *bel* is rather simpler in design and less expansive than the Bilari furnace. It is best suited to the needs of a small farmer.

It is also used in Bihar and is commonly referred to as *Improved South Bihar Furnace.*

Punjab

For the manufacture of *gur,* single pan furnaces are generally used in the Punjab State, though double or multiple pan furnaces are also used.

Improved Jalandhar Furnace

The furnace derives its name from Jalandhar where it was popular. This is an improved type of single pan furnace with arrangements for tilting the pan to overcome the danger of charring or caremalisation due to overheating. The furnace consists of two chambers separated by a grate. The chamber above the grate is 90 cm deep and serves as the combustion chamber while the bottom chamber (75 cm deep) serves as the ash-pit. A chimney (2.5 m in height) is provided for creating the necessary draught.

The boiling pan is flat bottomed with a diameter of 135 cm at top and 120 cm at the bottom with a depth of 26.4 cm. The furnace can boil 74.4 kg juice per hour and fuel consumption is about 54.1 per cent of the juice boiled.

Madhya Pradesh

In Madhya Pradesh, since 1925 old type of furnaces like *Malwa, Bhopal or Hadi Furnaces* have been used. Modifications have been made to improve thermal efficiency of these furnaces from time to time. These days *Improved Poona Furnace* and *Sindhwahi Furnace* are popular with the *gur* makers in this State (Sharma and Banger, 1991).

Karnataka

In the Karnataka region, an improved Hebbal furnace has been recommended for adoption.

The Improved Hebbal Furnace

The Agriculture Department, Karnataka, has designed this improved double pan furnace. One of the pans is flat bottomed and round with flaring edges. This is used as striking pan and is kept directly over the fire. The other pan is rectangular one placed at a relatively higher level than the striking pan and is usually heated by the flue gases. This pan can be directly heated when decolourisation of juice is needed for preparation of cream jaggery. A rectangular grating of 23 cm wide and about 44 cm long made of mild steel bars (1.35 cm in diameter) is fixed about 1.3 cm apart from each other at an angle in the furnace. This grating provides enough area to burn the bagasse completely. A door is provided on the grate for feeding the fuel. A 3.05 m high chimney made of plain galvanised sheet or masonary construction is provided to create sufficient draught to keep the furnace working satisfactorily. This furnace can boil five to six charges of 135.9 kg of juice (from 1.27-1.52 t cane) in a day of 12-14 working hours. Being efficient in fuel consumption, the bagasse of the cane crushed is usually sufficient for boiling the juice. The juice boiling pans are made locally of mild steel (or less commonly of copper).

Maharashtra

Different types of *Poona furnaces* are popular in Maharashtra and Karnataka.

Poona Single Pan Furnace (improved)

Being a smaller unit, it is suitable for cultivators having relatively smaller holdings. Usually the pan of this furnace has a capacity of 430-544 kg of juice per charge. With the introduction of high yielding varieties, it has been advantageous to increase the size of the pan in order tc boil 730-820 kg of the juice at a time. The furnace is built in two parts. The bottom, part is an ash pit 76.2 cm wide at the top, 121.92 cm at the bottom 121.92 cm deep and is always dug in the ground with an underground passage for admitting fresh air for burning the fuel. The furnace proper is constructed above the ash pit with a feeding hole of 45.72 cm wide and 91.44 cm high for fuel. The diameter of the furnace proper is 238.76 cm and its height is 111.76 cm. The air passage connecting the ash pit is 60.96 cm wide and 91.44 cm deep, opening outside. In the absence of grate, an iron sheet with a hole 25.4 cm sq in the centre is used. Two iron cross bars are fixed over it so as to avoid fuel falling through it into the ash pit, but at the same time allows the ash to drop through.

Juice boiling capacity of this furnace was further modified by increasing the diameter of the furnace (269.24 cm) and 127.0 cm height. The feed hole is 50.8 cm wide and 101.6 cm in height. The conical ash pit has top and bottom diameters of 86.36 cm and 137.16 cm, respectively. The depth of the ash pit is 137.16 cm. The air passage connecting ash pit is about 68.58 cm wide and 99.06 cm deep opening out side. The construction of the iron grating of the furnace is similar to that of *Poona single pan (improved) furnace.* This furnace is known as *Poona single pan furnace* (modified).

Poona Double Pan Furnace

This furnace is a combination of two furnaces as described above. Both the furnaces are connected by means of a flue pipe, constructed at a higher level. A slanting flue pipe (68.58 cm wide and 34.24 cm high) rises from the lower furnace and opens into the upper furnace. In front of this opening, a circular baffle wall (31.1 cm high and 12.7 cm wide) at the top, is placed in the upper furnace with a view to raising and spreading hot gases emanating from the lower boiling furnace. The horizontal distance between the centre of the baffle wall at the sides of the furnace is 96.52 cm. The space enclosed by the circular baffle wall on the sides of the boiling furnace is made like a saucer or half saucer. On the outside the space between top of the wall and the lower side of the opening towards the chimney is filled to form a slope from edges of the circular ridge to the chimney which is 30.48 cm in diameter and 3.35 m in height. The distance between the base of the ridge at the bottom of the pan i.e., the depth of upper furnace is 68.58 cm while the diameter of the upper furnace is 269.24 cm. The horizontal distance between the two furnaces is about 101.6 cm while vertical distance between the upper edges of the two furnaces is about 61 cm.

A storage pan is provided with a stop cock, kept on the upper furnace. Juice from the storage pan is tapped into an open trough, and then led to the boiling pan kept on the furnace. Initially, both the pans are filled with juice (730-820 kg) and as the boiling pan is struck and emptied, it draws upon more juice from storage pan, while the latter gets its replenishment of the fresh juice.

Poona Triple Pan Furnace

The triple pan furnace consists of two single pan type furnaces, one for boiling juice and the other as storage pan, placed in triangle. Its construction and capacity are almost similar to that of the *Poona double pan furnace.*

Besides this triple pan furnace, there is a still bigger furnace called *Poona Multiple Furnace* which is exactly similar in design to the triple pan furnace excepting that another additional pan

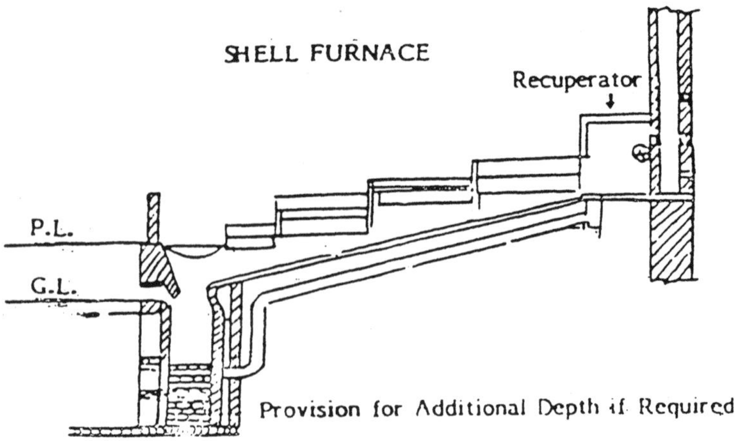

Fig. 15.2: Schematic Diagram of the Shell Furnace

is placed above the third pan for further utilisation of the hot flue gases.

Selection of the desired type of furnace mainly depends on the holdings of cane acreage to be disposed each day. The Poona double, triple or multiple pan furnaces are generally used for large scale manufacture of *gur,* as capacity of the boiling pans of these furnaces are 730 to 820 kg.

Andhra Pradesh

Anakapalle Furnace (improved)

In Andhra Pradesh, at Anakapalle, double grating Anakapalle furnace was designed in 1945 by making suitable modification in the then popular *Godavari furnace,* on the line of *Sindhwahi furnace.* The modified furnace was found satisfactory both from the point of fuel economy and efficiency. The fuel consumption in this furnace was observed to be 35 to 39 per cent of dry air bagasse or 45 per cent of trash on weight of juice. The average time taken in March for boiling down a charge of 360 kg of juice was about 2 hours. Feeding of the furnace was noted to be easy and steady boiling of the juice could be ensured in this furnace.

For boiling sugarcane juice to *gur* wide shallow pan of about 2.25 m diameter and 46 m depth were found to be most efficient. Among various pans used, concentric ring riveted pans was observed to be sturdier and lasted relatively longer than the cut-sheet metal pans (Anon. (Ankapalle, India), 1989).

Improvements in Furnace Efficiency

Shell Furnace

In Open Pan Sugar (OPS) technology one of the major advances is the evolution of *Shell Furnace* (Fig. 15.2) for wet bagasse burning. With this it is possible to increase the sugar recovery by a further 0.8 to 0.9 per cent and thus it is possible to obtain a recovery of 8.3 per cent from the existing level of 7.5 per cent (Varma and Garg, 1987).

Biomass Residue Gasifier

Normally, furnaces are very inefficient (10-15% overall efficiency). At the Nimbkar Agricultural Research Institute, Phaltan (Maharashtra), studies have been conducted to develop a furnace coupled with a low density biomass gasifier. This multi-fuel gasifier can run on sweet sorghum bagasse, sugarcane trash, wheat husk, grasses, etc. (Rajvanshi and Jorapur, 1992). This gasification system has been used for *gur* making. The pilot gasification plant is of 100 kW (thermal) capacity and produces about 15-20 kg of sweet sorghum *gur* per batch. The advantages of this gasifier over the conventional furnace are as follows:

(i) a fully automatic system,
(ii) produces a bluish-white flame with a flame temperature in excess of 1250°C, relatively smokeless,
(iii) overall fuel efficiency 20-27 per cent (w/w) and
(iv) gasification system produces char as by-product (20-25% w/w of fuel) which can be mixed with a suitable

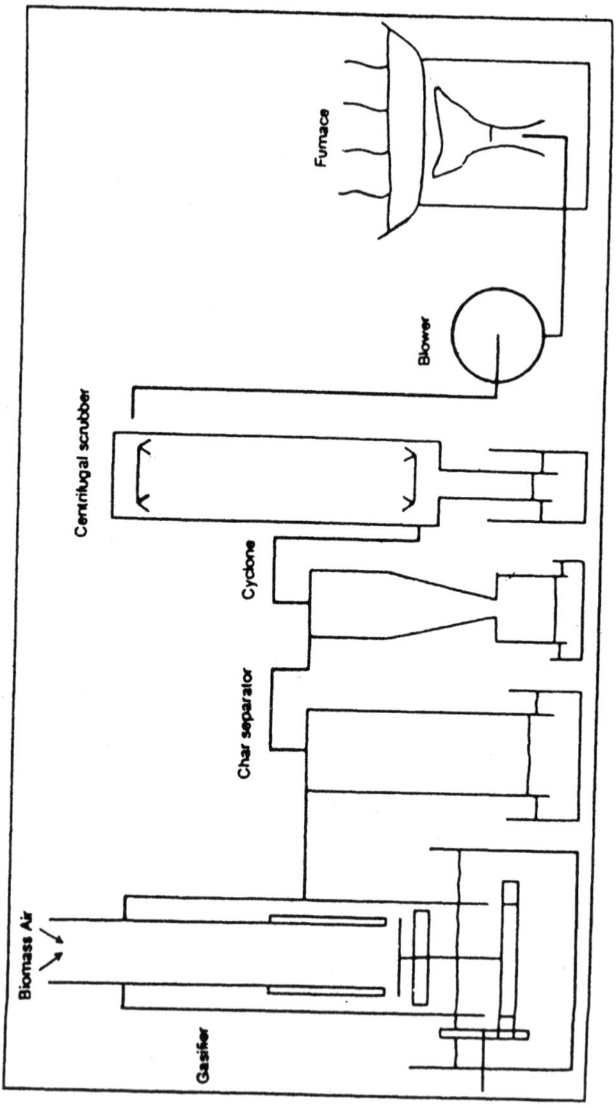

Fig. 15.3: Schematic Diagram of Biomass Gasifier-powered Jaggery or Syrup-making

binder to make excellent smokeless fuel briquettes for domestic use.

A systematic presentation of gasifier-powered *gur* or syrup making furnace is shown in Fig. 15.3. Rajvanshi *et al.* (1993) have advocated use of this biomass gasifier for making *gur* or syrup from sweet sorghum.

Fuel for Heating: The Bagasse

Bagasse, an easily available fibrous material having low ash content, is obtained by crushing cane and is used as fuel for firing furnace(s) for boiling sugarcane juice to the desired level of concentration (or the striking temperature) for making *gur*. Bagasse contains 48-52 per cent moisture, 1.7 per cent ash, 2.5 per cent sucrose and other soluble impurities. Its gross calorific value is 19400 KJ/kg (Upadhiaya, 1991). It contains 50 per cent moisture which escapes in the furnace and gets into the flue gases, carrying with it the latent heat and sensible heat. If wet bagasse is fed to fire the furnace, the combustion zone gets extinguished (Varma and Garg, 1987). By drying bagasse, besides more efficient burning, the losses (as mentioned earlier) are also reduced and contribute to fuel efficiency of the furnace. This also results in saving of some bagasse which may be utilised otherwise also. The net calorific value (btu/lb bagasse) is 7607, 7171, 4988, 4115 and 3242 for a bagasse containing 0, 5, 30, 40 and 50 per cent of moisture, respectively (Doss, 1987). Normally the bagasse is dried in the sun near and around the crushing yard and/or the furnace.

REFERENCES

Anonymous, (1955). An improved furnace for the manufacture of gur (jaggery): Bilari bel No. 1. *I.I.S.R. Newsl.*, I(1-12): 18-20.

Anonymous, (Anakapalle, India) (1989). *75 Years of Agricultural Research at Anakapalle: 1913-1988*, RARS, Anakapalle, A.P.A.U., Anakapalle, A.P., pp. 400.

Doss, K.S.G. (1987). Drying of bagasse by a new concept (Patent applied for), in *Platinum Jubilee Souvenir* Vol. I, Sugarcane Breeding Institute, Coimbatore, India, pp. 170-174.

Rajvanshi, A.K. and R.M. Jorapur (1992). *Development of Leafy Biomass Gasification Systems, Publication No. NARI-GS-1*, Nimbkar Agricultural

Research Institute, Phaltan, Maharashtra, India.

Rajvanshi, A.K., T.K. De, R.M. Jorapur and N. Nimbkar (1993). *Jaggery and Syrup from Sweet Sorghum*, Publication No. NARI-GUR (1993), Nimbkar Agricultural Research Institute, Phaltan, Maharashtra, India, pp. 10.

Sharma, S.R. and K.S. Bangar (1991). Status of *gur* making furnace in Madhya Pradesh, presented at the *National Seminar on Increasing Productivity, Quality and Storability of Jaggery (Gur)* organised by the Indian Institute of Sugarcane Research, Lucknow and the Indian Society for Agricultural Engineers, New Delhi at the Indian Institute of Sugarcane Research, Lucknow on Nov., 30-Dec., 1, 1991.

Varma, N.C. and M.G. Garg (1987). Recent advances in the open pan sulphitation process of cane sugar manufacture. In: *Bharatiya Sugar Directory*, pp. 258-268.

Upadhiaya, U.C. (1991). Bagasse as a fuel. *Intern. Sugar J.*, 93(1111): 132-138.

16

MEDICINAL USES OF *GUR*

India is the major sugarcane producing country in the world and approximately 54 per cent of the produce is utilised for the manufacture of *gur,* the age old nutritive sweetener known to mankind. In Uttar Pradesh (the major sugarcane producing state), 68 per cent (Anon., (Shahjahanpur, India), 1983) and in Andhra Pradesh 60 per cent (Chiranjeevi, 1987) of the sugarcane produced is utilised for *gur* manufacture. *Gur* is also prepared from the sugary sap obtained by tapping either the stalk or the inflorescence of certain palms, or by crushing sweet sorghum stalks. *Gur* is an open-pan nutritive sweetener which is easily available to every one in the country.

Medicinal value of sugarcane and *gur,* and products derived from them, have been recognised by our ancestors and medicine men from times immemorial. They have advocated their use in maintaining a heathy society. Regarding chewing of sugarcane, *Charak,* a well-known ancient Indian Ayurvedic medicine man, opines that:

बृष्यः शीतः स्थिरः स्निग्धो
वृहणो मधुरो रसः ।
श्लेष्मलो मक्षितस्येक्षो-
र्यान्त्रिकस्तु विदह्यते। ।
{(चरक (सूत्रस्थान, अ० 27, श्लोक 234)}

(the juice of chewed cane is heavy, but gives soothing and cooling effect besides increasing semen. It also induces *cough* (phelgm) formation. The crusher extracted juice is *vidahi*).

Shushruta, another great Indian medicine man, opines that:

अविदाही कफकरो वातपित्त निवारणः ।
वक्त्र प्रह्लादनो वृष्णो दन्त मिष्पाडितो रसः ।।
{सुश्रुत (सूत्र-स्थान अ 45, श्लोक 157)}

(that chewed juice is *avidahi, cough* (phelgm) inducing, eliminates *vata* (air) and *pitta* (bile). The juice gives pleasing effect and concomitantly increases semen).

The *Ashtang Sangrah,* an authoritative Ayurvedic book, explains several good qualities of sugarcane juice as mentioned below:

इक्षुरसो गुरूः स्निग्धो बृहण कफमूत्र कृत् ।
वृष्यः शीतः पवन जिद् मुक्ते वात प्रकोपेनः ।।
रक्तः पित्त प्रशभनः स्वादु- पाक रसः सरः ।
सोऽग्रे सलवेण दन्त पीडितः शर्करा समः ।।
{अष्टांग संग्रह (सूत्र-स्थान द्रव द्रव्य विज्ञाकीय अध्याय 6,
श्लोक 81, 82)}

(the juice of sugarcane is heavy, *cough* (phelgm) inducing, refines urine, improves sex-potency, cools semen, and eliminates *vata* (air). But when taken after meals, it makes *vaau* aggressive. It pacifies blood and *pitta*).

Gur as Medicine

In Ayurveda, *gur* has been considered as a nutritionally rich substance endowed with innumerable medicinal qualities. *Gur* has been called *rasayan* i.e., which extends the life, keeps the body disease free and maintains *yauvana*- the bloom of youth. *Bhava Prakash* another book on Ayurvedic medicine-advocates that the use of *gur* with ginger pacifies *cough,* with *harr* eliminates *pitta* and with *sonth* (dried ginger) is beneficial in all the diseases caused by *vata* factor. *Harita Samhita* considers *gur* as the best eatable in *kshya* (consumptive diseases), bronchitis, coryza, weakness and anemia, jaundice, etc. As per *Rajanighuntukara, gur* is good for cardiovascular system, pacifies *vata, pitta* and cough; eliminates the diseases of urinary tract and bowels; invigorates, tones and improves digestive power; and cures itching and *prameha*- the diseases in which sperms and other vital fluids of the body pass out through urine.

As per *Shushruta Samhita*

पित्तघ्नो मधुरः शुद्धो वात घ्नोऽसृक्प्रसादन ।
स पुराणोऽधिक गुड़, पथ्यतमः स्मृतः ।।
(सुश्रुत संहिता, सूत्र स्थान, अध्याय 45)

(pure *gur* eliminates *vata* and *pitta*, purifies the blood. Old *gur* is particularly beneficial, when given to a lactating mother after child birth).

On the Ayurveda Day Celebrations at the Regional Ayurveda Sansthan, Lucknow, on October 24, 1992, it was advocated time and again that presence of beneficial minerals like manganese and magnesium increases the medicinal value of *gur* and its regular use considerably prolongs the life span of man (*Rashtriya Sahara*—a Hindi daily). For elderly persons, use of *gur*, in place of sugar is considered an important dietary component of the food. In eastern U.P., an old sayings is:

चैत-चना, बैशाखे-बेल,
जेठे-शयन, आषाढ़े-खेल,
सावन-हर्रे, भादों-ती सी,
क्वार मास गुड़ खाओ मीतः ।
कातिक-मूली, अंगहन-गाजर,
पूषमाघ-करो दूध से मेल,
माघमास घी खीचड़ खाय,
फागुन उठते प्रात नहाय ।।
(*साभार:* श्री सूर्यनाथ लाल श्रीवास्तव, ग्राम-हरपुर, बलिया, उ. प्र.)

(It indicates that the use of *gur* in *kwar* (in October) is beneficial for health. *Gur* is generally consumed during winter in several preparations made in combination with *til* (*Linum* sp.), and ground nut kernel. These preparations are sold in the market in the name of *revari, gazak chikki, etc.*)

Gur in Curing Diseases

Use of *gur* has been known to cure certain diseases. Some of the uses of sugarcane and *gur* for the treatment of human diseases advocated by Ayurveda have been mentioned by Gaur (1974) and Lakshman (1988). Relevant information is collated and presented below:

Coryza

Use of small quantity of the decoction of *sonth* (dried ginger), turmeric and *gur* provides an effective prophylactic measure against coryza (Soni, 1991).

Coryza can also be cured by taking the following:

(a) A pellet of *gur,* containing powdered asafoetida and 2-4 black peppers (placed in the centre of the pellet), in the morning and evening is quite useful in controlling coryza (Soni, 1991).

(b) Chewing 1-2 pellets of pea size, thrice a day containing equal quantity of *sonth, gur* and *mulhethi (Glycerrhiza)* provides an effective control of coryza.

(c) For curing spoiled coryza, 10 g *gur,* 40 g curd and 2.5 g of powdered black pepper should be taken daily in the morning for 3 days. Besides, this preparation also cures foul smell emanating from mouth as well as hoarseness in the throat.

Respiratory Ailments

Taking 2.5 to 10 g *gur* with equal quantity of pure mustard oil daily for a period of 21 days cures respiratory ailments. To stop hiccup, a few drops of *gur* solution in water having ground *sonth* should be dropped in the nostrils.

Palm gur is also used in curing coryza and pulmonary ailments.

Headache

Ten gram *gur* is ground well with 5 g of black *til* (linseed) and mixed with milk. The 5 g of *ghee* (preferably made from cow-milk) is mixed with it and the material is heated. This preparation (lukewarm) is applied externally on the forehead and sides to cure headache.

Rheumatism

Administering orally a preparation of *gur* and cumin is beneficial in curing rheumatic diseases. Mixing *gur* with fresh milk of buffalo (to sweetness) and taking it while standing is beneficial. Besides, remaining standing for 2-3 hr (do not sit) is beneficial in *nala-vata.* For rheumatic pain in knee joints administering pellets of finely mixed *guggul* (10 g) and *gur* (20 g) with some *ghee* in the morning and evening is beneficial.

Cardiac Diseases

Taking equal quantity of *gur* and pure *ghrita* regularly is beneficial in curing cardiac diseases. Eating anything made up of *gur* has its own beneficial effect in such cases.

Disorders of Digestive System

Taking *gur* with sweet *bel* powder is beneficial in almost all the stomach disorders. For diarrhoea, mix ginger or *sonth* (the dried ginger) or *harr or peepal* powder; start with 10 g of this preparation daily and increase up to 120 or 150 g daily. Besides, diarrhoea, this preparation also cures coryza, hoarseness, respiratory troubles, sickness of stomach (anorexia), chronic fever and piles. Taking mixture of 30 g pulp of unripe *bel,* 5 g *sonth* (the dried ginger) and 10 g *gur,* twice or thrice daily is also beneficial in diarrhoea.

In case of loss of appetite, taking 5 g *sonth* with 15 g of *gur* twice or thrice overcomes the ailment.

In *raktatisar* (diarrhoea with bleeding) taking cooked pulp of unripe *bel* with *gur* is beneficial.

Piles

Administering *gur* and *harr* powder, orally, cures piles. As per another prescription, administering *choti harr* grinded in cow urine with *gur* in the morning also cures it.

Inflammation or Burning Sensation in the Body

Dissolve *gur* in water, filter it 25 times through a clean cloth, Drinking this for many times a day cures body inflammation.

Kidney Pain

Ten gram of *gur* is mixed well with 0.5 g slacked lime to make two lobules. To relieve kidney pain, lobule should be taken with luke warm water and if the pain does not subside, the other lobule can also taken.

For relieving pain after the stones in kidney or ureter have come out, patient should sit in warm water (tolerable) and should drink warm milk with *gur*.

For Worms

For controlling worm infection, make pellets having 3 g *ajawain*, 3g *byvedang*, 0.1 g *kapoor* and 5 g *gur*. These pellets should be taken orally 2 to 3 times a day to kill worms. It has also been observed that tablets having only 0.5 g *ajawain* and 0.5 g *gur* when taken thrice in a day also kills the worms.

Leucorrhoea

Taking daily 5 g of the ash prepared by burning gunny bag in which *gur* was stored at least for 3 years has been found to effectively cure leucorrhoea. This preparation is also beneficial in excessive bleeding in menstrual cycle.

Production of Relatively Less Semen and Impotency

Regular consumption of *gur* with *aonla (Phycus ambelica)* powder increases the quantity of semen. This preparation also cures *rakta-pitta*. Besides, it is also beneficial in *mutra-kricha*. Powder of *tulsi (Ocimum* sp.*)* seeds or the roots with some quantity of old *gur* when taken 1.5 to 3 g daily with cow milk for 5 to 6 weeks also helps in overcoming impotency.

Family Planning

If a lady, after the menstrual cycle, takes 40 g *gur* with warm water daily for a period of 15 days, pregnancy can be avoided. As per *Kuchimartantra,* this prescription is helpful in planning the family.

A preparation containing *Chirchitta* [*Cenchrus biflorus* Roxb. (Poaceae)] seed powder mixed with onion juice and three-year old *gur* is quite effective for inducing abortion in ladies (Sharma, 1991).

Thorn, Glass and Stone Injury

Application of melted *gur* (tolerably warm) on wounds caused by thorns, glass or pointed stone (which has pierced through flesh), relieves the pain and quickly heals the wound.

If a thorn, pierced deeper in the flesh, is not coming out easily, then a warm poultice of *gur* and *ajawain* is placed over the injured place. After a few hours thorn comes out automatically (Jain, 1992).

Hair Care

Applying the paste made up of overnight soaked Trigonella (*methi*) seeds and *gur* to hair on the scalp and washing the hair after one hour increases the lustre of hair. At the same time it overcomes the dryness of the hair (Bharadwaj, 1993).

Waxing is done to remove unwanted hair, especially in ladies. Wax may be produced domestically using *gur* as a base material. For this, a cupful of *gur* is mixed with a cup of water and two teaspoonful of lemon juice. It is heated on low flame with intermittent stirring. When the solution turns dark brown it is removed from flame and cooled. Half teaspoon of glycerine is mixed for waxing.

Administering juice of *Cucumber pepo (kaddu)* mixed with *gur* is useful in bringing a drunkard person to senses (Anon., 1991).

Besides being used as a medicine, *gur* is also used as carrier of medicines.

These medicinal uses of *gur* have been compiled from various demi-scientific sources and popular magazines and news papers. It is advised that, these preparations, though unharmful, should be administered only under medical supervision.

Gur in Veterinary Medicine

Preparations involving *gur* also find their use in veterinary medicine. Boiled leaves of *goma (Leucas urticaefolius* (Laminaceae) mixed with *gur are* given to cows and buffaloes for expelling placenta after delivery (Sharma, 1991).

REFERENCES

Anonymous (1983). Shahjahanpur, India.

Anonymous (1991). *Kumhra. Ayrogya Vardhini,* p. 37.

Bharadwaj, V. (1993). *Baten kam ki. Rashtriya Sahara* (Hindi Daily): Feb., 17.

Gaur, G.P. (1974). *Doctor Ganna,* Tej Kumar Book Depot. (Pvt.) Ltd., Lucknow, p. 64.

Jain, A. (1992). *Kuch Gharelu Nuskhe. Rashtriya Sahara* (Hindi Daily), July 8, p. 14.

Lakshman, H. (1988). Use of sugarcane, sugar and jaggey in the treatment of human diseases in Ayurveda. *Platinum Jubilee Souvenir* RARS, Anakapalle, Andhra Pradesh, pp. 139-142.

Sharma, M.P. (1991). Less-known medicinal uses of plants from Mewat (District Gurgaon), Haryana, India, *Econ, Bot.,* 45: 435-436.

Soni, L. (1991). *Ayrogya Vardhini,* p. 86.

17

RESEARCH NEEDS IN GUR

Gur or non-centrifugal sugar is made in different forms in various parts of the world. Several aspects of the lump sugar, as reviewed in the foregoing 16 chapters, indicate that appreciable amount of progress has already been made. However, to make *gur* industry more lucrative and attractive, and to ensure regular supply of quality *gur* to the consumers, certain aspects still need a sustained research endeavour. Some of these are mentioned below:

Mechanism of Clarification

Clarification is an important step in *gur* manufacture. Studies are needed to elucidate mechanism of clarification by studying the process of interaction between juice colloids and clarificants of plant origin. Such studies may help in identifying or developing clarificants with optimum efficacy.

Claribility of Juice

Clarification of sugarcane juice is influenced by cultural practices (*e.g.*, manuring, time of planting, harvest, spacing, irrigation, *etc.*) and ingredients in the juice like colloidal substances, non-protein N and minerals. More studies are required to work out interaction between variety, soil, fertilisers and cultural practices with reference to their bearing on ingredients which affect claribility of juice and quality of the end product, the *gur*. Studies on certain cultural practices have shown that an inherently poor *gur* variety may yield a superior quality *gur* at a nitrogen level of 150 kg/ha with 6 irrigations during summer as against inferior quality of *gur* at 75 kg/ha with 2 irrigations in hot summer months. These could decide optimal cultural conditions so as to derive maximum benefits from the best variety or for that matter any variety in *gur* areas. In this regard, commendable efforts have been made by the scientists of the Sugarcane Research Station, Anakapalle, Andhra Pradesh, but to improve upon the *gur* quality and quantity, more work is needed in other *gur* areas.

Microstructure of *Gur*

Except for the studies of Varahalu (1938, 1939), efforts have not been made to study the microstructure of *gur, rab, etc.* More research efforts are needed to elucidate the microscopic structure (preferably by using electron microscope) as it may provide basis for differentiating good and poor quality *gur* and may provide clue(s) to the better storability (and deterioration) during storage. This may pin point the structural component(s) responsible for retaining and/or deteriorating quality of *gur*. Microstructure of *gur* made by using different clarificants will also be desirable to attain or alleviate certain structural components contributing to *gur* quality.

Colour of *Gur*

Interaction of iron and phenolics primarily determines the colour of *gur*. Study of such an interaction will be desirable in varieties producing light or dark coloured *gur*.

Aging Studies in *Gur*

Medicinal properties of old *gur* have been well documented. Long term storage of *gur vis-a-vis* its therapeutic value, appears to be the need of the hour. In honey, compositional changes with time lead to darkening of colour, loss of aroma and flavour. Long term effects of heat and storage result in an increase in hydroxy methyl furfural content to a greater degree; however, the diastase number appears to be a limiting factor for freshness in short term storage. Therefore, these parameters are used to evaluate the freshness of honey and these together can be used to judge its processing and conditions for storage (Sancho *et al.*, 1992). There is need to initiate such type of studies in *gur* also to delineate certain criteria for freshness and proper storage.

Residue Analysis in *Gur*

Raising a good sugar crop without the use of pesticides (herbicides, fungicides, insecticides, *etc.*) is rather difficult though not impossible. These chemicals may leave undesirable residues in *gur* prepared from the juice of treated canes. Their assessment is important not only from the point of view of basic information but also from human health view point as the per capita consumption of *gur* and *khandsari* by 2000 A.D. is likely to be around 11.5 kg/annum (Anon., 1976).

Efficient Furnace for Making *Gur*

An efficient furnace will be a boon to the *gur* manufacturers. Particularly one that economises on fuel consumption will be desirable. *Shell furnace* and *biomass gasifiers* appear to have promise in this regard. There is also a genuine need to develop furnaces based on solar heat.

Gur from Sugar Cane Raised Under High Levels of Management and the Crop Sprayed with Chemical Ripeners

Though some work has been done on these lines but more is needed to be done. Obtaining high yields under good management using higher levels of input is the need of hour. The cane so obtained may pose some problems in juice quality and recovery of *gur, etc.*

Under certain conditions when the crop continues to grow vegetatively and fails to mature, ripening is induced either by withholding irrigation or by applying certain chemicals commonly referred to as chemical ripeners. These include Polaris, Polado, Aseulaux, Ethrel, sodium metasilicate, Glyphosate, *etc*. They improve sugar content and increase sugar output. However, practically no information is available with regards to ripeners on the recovery of *gur,* its quality and behaviour during storage. Presence of residues of these chemical ripeners in *gur* should also be investigated using modern methodology, particularly gas liquid chromatography, *etc.*

Gur from Pest and Disease-affected Canes and Problem Soils

It has been unquestionably documented umpteen times that *gur* does not set well when juice is obtained from canes grown on problem soils, such as, waterlogged or saline ones. There is a pressing need to evolve a process whereby the excess of sodium and potassium can be effectively reduced in juice particularly from canes grown on such soils. Similar difficulties are experienced with pest and disease ridden canes. Blanching with hot water has been developed as one of the technologies for improving *gur* recovery and quality by the scientists of the Regional Research Station, Anakapalle, to make *gur* from such canes (Anon., (Anakapalle, India), 1989). But more research is needed on this aspect.

Gur in Human Health

Most of the information regarding the use of *gur* in relation to human health and curing of certain diseases is based on the ancient Ayurvedic literature and scattered rural knowhow. *Gur* contains most of ingredients of a wholesome diet which are a must for sustaining good health. Recently it has been shown, but not conclusively demonstrated, that daily use of *gur* increases the life span of a man atleast by seven years. It has also been frequently observed that the incidence of diabetic patients in urban areas (sugar consuming) is more as compared to rural (*gur* consuming). The scientific reason for this should be investigated. Old *gur* has been claimed to be of superior medicinal value as compared to the new one. Constituents responsible for this should be investigated. Further, medical practitioners should come forward to recommend *gur* for human health and for curing those diseases for which its efficacy has been proved beyond doubt. *Gur* prepared from cane growing in limiting environments like waterlogging, saline and frost prone conditions, and pest and disease affected canes may have some medicinal value this needs to be explored on scientific lines.

REFERENCES

Anonymous, (1976). Rept. *National Commission on Agriculture* (Pt. III), *Demand and Supply*, Govt. on India, Ministry of Agriculture and Irrigation, New Delhi, pp. 195.

Anonymous, (Anakapalle, India) (1989). *75 Years of Agricultural Research (1913-1988) at Anakapalle,* Regional Research Station, Anakapalle, A.P.A.U., Hyderabad, Andhra Pradesh, pp. 400.

Sancho, M.T., S. Muniategui, J.F. Huidobros and J.S. Lozano (1992). Aging of honey. *J. Agric. Food Chem.,* 40: 134-138.

Varahalu, T. (1938). Studies in sugarcane jaggery. V. Macrostructure of *jaggery*. *Madras Agril. J.,* 26(5): 394-398.

Varahalu, T. (1939). Studies in sugarcane jaggery. VI. Microstructure of *jaggery*. *Proc. Annl. Conv. S.T.A.I.,* 8: 1-20.

SUBJECT INDEX